WHY NOT
NONLINEAR ALGEBRA
Math, Science, Applications and Funding

WHY NOT NONLINEAR ALGEBRA

Math, Science, Applications and Funding

Michael F Shlesinger

NEW JERSEY · LONDON · SINGAPORE · BEIJING · SHANGHAI · HONG KONG · TAIPEI · CHENNAI · TOKYO

Published by

World Scientific Publishing Co. Pte. Ltd.

5 Toh Tuck Link, Singapore 596224

USA office: 27 Warren Street, Suite 401-402, Hackensack, NJ 07601

UK office: 57 Shelton Street, Covent Garden, London WC2H 9HE

Library of Congress Control Number: 2025042245

British Library Cataloguing-in-Publication Data
A catalogue record for this book is available from the British Library.

WHY NOT NONLINEAR ALGEBRA
Math, Science, Applications and Funding

Disclaimer: The views and opinions expressed in this book are those of the author(s) and do not necessarily reflect the views of the publisher. While every effort has been made to ensure the accuracy of the information contained herein, the publisher assumes no responsibility for errors or omissions, or for any consequences arising from the use of the information provided.

ISBN 978-981-98-1462-6 (hardcover)
ISBN 978-981-98-1573-9 (paperback)
ISBN 978-981-98-1463-3 (ebook for institutions)
ISBN 978-981-98-1464-0 (ebook for individuals)

For any available supplementary material, please visit
https://www.worldscientific.com/worldscibooks/10.1142/14352#t=suppl

Desk Editors: Soundararajan Raghuraman/Lai Fun Kwong

Typeset by Stallion Press
Email: enquiries@stallionpress.com

Dedicated to my family

*Nava, Adi, Josh, Jordan, Leo, Jacob, Ben, Sara, Ellie, Natalie,
Dan, Michelle, Emma, Dylan*

Preface

In 1840, Richard Dana wrote *Two Years Before the Mast* about his time, leaving Harvard, for a break from student life, to work on a sailing ship plying the leather trade up and down the California coast, which was then part of Mexico. Before the mast is a nautical term designating being an ordinary seaman quartered in the forecastle. Dana Point on the California coast, south of Los Angeles, is named for him. I borrow the term "mast" to describe my "nautical" 40 years as an ordinary Program Officer at the Office of Naval Research, quartered at home.

Due to the 1849 California Gold Rush, Dana's book became sought after by the 49'er gold miners for its valuable information about California. The only valuable information I can provide in this book, from an ONR perspective, is that grant funding will come naturally when you follow your interests, enjoy your research, make friends, and do good work.

About the Author

Michael F. Shlesinger received his BS in Math and Physics from Stony Brook University and his PhD in Physics from the University of Rochester. He has worked at the University of Rochester, Georgia Institute of Technology, the University of Maryland, the La Jolla Institute, and for over 40 years at the Office of Naval Research. At ONR, he headed the Physics Division, was a member of the Senior Executive Service, a Chief Scientist for Nonlinear Physics with a Presidential Rank Award, and received ONR's Saalfeld Award for Outstanding Lifetime Achievement in Science. In 2024, he was awarded the Victor Weisskopf Medal for Outstanding Scientific Statesmanship. He has held the Kinnear Professorship at the USNA. He is a Fellow of the American Physical Society with their Outstanding Referee Award. He co-founded the World Scientific journal, *Fractals,*

and was a Divisional Associate Editor of the *Physical Review Letters*. He co-founded the Experimental Chaos Conference. His research has twice been the cover article in *Physics Today*, and his research in statistical physics can be found in his book, *An Unbounded Experience in Random Walks with Applications*. He has given the Michelson Lecture at the USNA and the Regents' Lecture at UCSD. While his works are mainly in statistical physics, he has patents on protein design.

Contents

BLUF

Naval officer reports start with BLUF, "bottom line up front", so I'll do the same. This book is about my time at the Office of Naval Research, how I came for 1 year to learn the grant business from the inside with the plan to return to my university position. ONR turned out to be much more than reading proposals and awarding grants. It was more about creating and defending programs of exceptional science with national defense implications. I could join teams with great colleagues to create multidisciplinary programs. To my surprise, the Navy afforded me so many interesting opportunities, travels, and support, plus time for my own scientific career. Every day was different and interesting. Some occasions were briefing the Secretary of Defense on chaos theory, discussing crane control with the Chief of Naval Operations, meeting with the Under Secretary of the Navy's three-star group, and speaking with the Chairman of the House Armed Services committee about nonlinear sensors, and I toured China's Dun Huang caves, had M4 rifle practice at the Quantico

Marine Corps base, discussed complexity theory with the Lt. General in command of the Marine Corps Combat Development Command, helicoptered out to the USS Kearsarge, interacted with NCIS, was the Blue National Command Authority at the Naval War College Sea Conflict War Games, co-authored a report on labs for the Defense Science Board, co-authored a Foreign Assessment report on Soviet Union research on nonlinear dynamics for SAIC, and co-created a journal *FRACTALS* and an international conference on Experimental Chaos. I had a Kinnear Professorship at the USNA. Joyfully, I ended up staying at ONR for 40 years. This is that story.

CHAPTER 2

A Prophetic Remark

At the State University of New York at Stony Brook, in 1968, I was a math major taking a course in linear algebra. The new department chair was James Simons, and he held a department tea. He asked me what I wanted to study. Being clueless, I said multilinear algebra, which I thought would be the extension of my present course. He gave me a puzzled look and said, why not nonlinear algebra, the title I've used for this book. I didn't know what that meant and couldn't have prophesied that I would spend 40 years at the Office of Naval Research managing a nonlinear science program that indeed included nonlinear algebra. I almost called this book after the children's game, Simon Says. By the way, Simons left Stony Brook and founded the famously successful hedge fund, the Renaissance Fund. What follows are my stream of consciousness thoughts on nonlinear mathematics and science.

Why Math and Physics

The payoff from basic research has been ever-progressing but with a time delay. It was well said by Frank Wilczek in Science (March 1, 1991), "Basic research of any kind requires patience and sympathy. Progress is often fitful, and the value of a really new idea may take years to appreciate and may ultimately prove itself in totally unexpected ways". I used to quote this when giving ONR presentations. Did our predecessors think about the payoff? Newton invented calculus to derive Kepler's laws for an elliptic orbit for the Moon. Was there a payoff? Did anyone think of an application? Going further back, Euclid proved that there is an infinite number of prime numbers. Any payoff or applications? Still, these great scientists and mathematicians lit the spark that brought us to today. I guess it was just the love of learning and following one's curiosity. Sometimes at ONR, the question is payoff. I guess my answer has been, ask Euclid and Newton.

I assume mathematics developed with the need for counting and that led to arithmetic. Geometry probably followed for determining lengths and areas needed for construction materials. Eratosthenes, around 240 BC, even measured the size of the Earth through simple geometry. What a brilliant thought, but did anyone take advantage of this calculation? What was the payoff to go beyond these beginnings? What would early mathematicians guess? But the payoff has been enormous with mathematics being the foundation of theoretical physics. I guess bright people were drawn to puzzles and kept pushing the boundaries of mathematical creation and knowledge. I think today's mathematicians feel a kinship with them.

Here's a way a measurement led to an advance in mathematics. If you measure the two sides of a right triangle, how can you determine the hypotenuse? This is the famous Pythagorean theorem. If you have forgotten, how to prove it, go from the triangle to a square comprised of four triangles and an interior square, so it's not thinking outside the box but constructing a box.

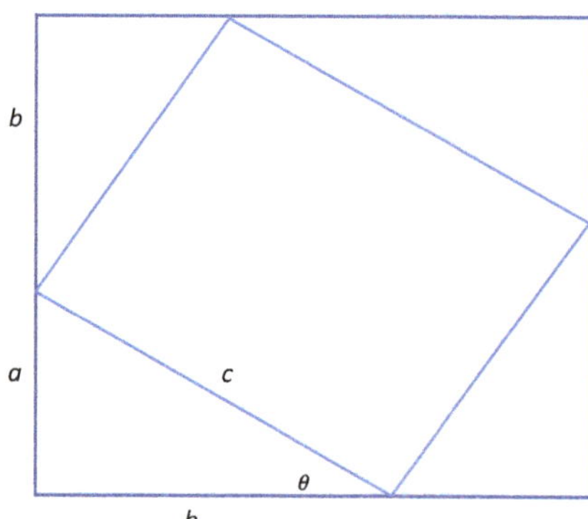

The square which has an area $(a + b)^2 = a^2 + 2ab + b^2$ and breaking that down to the four triangles and internal square which has an area equal to $4\left(\frac{1}{2}ab\right) + c^2 = a^2 + 2ab + b^2$; thus, $a^2 + b^2 = c^2$.

Rewriting the above as $\left(\frac{a}{c}\right)^2 + \left(\frac{b}{c}\right)^2 = 1$ and with θ the angle at the bc vertex, then writing $\frac{a}{c} = \sin\theta$ and $\frac{b}{c} = \cos\theta$, yielding $\sin^2\theta + \cos^2\theta = 1$, voila the birth of plane trigonometry.

Algebra should arise when in an equation one of the variables is unknown and must be solved for. This led to the puzzle for quadratic equations that some solutions had the square root of a negative number. That seemed to make no sense. In elementary school, when teaching multiplication, it is stated that $1 \times (-1) = -1$ and leaves it at that as a rule. What if it is taught this way? Thinking outside the box, by going to two dimensions (the complex plane) is another way of introducing an imaginary number. Go to two dimensions and rotate counterclockwise through $180°$ to go from 1 to -1.

One can also do this in two stages, by rotating by $90°$ twice. Call the first rotation i and two rotations in a row $i^2 = -1$. Of course, engineers would use j instead of i. So, rotations in a plane make a complex number appear naturally. Rotations in three dimensions bring one to quaternions with three types of complex numbers i, j, and k, but no need to go into that here. What seemed strange in solving quadratic equations becomes essential in physics.

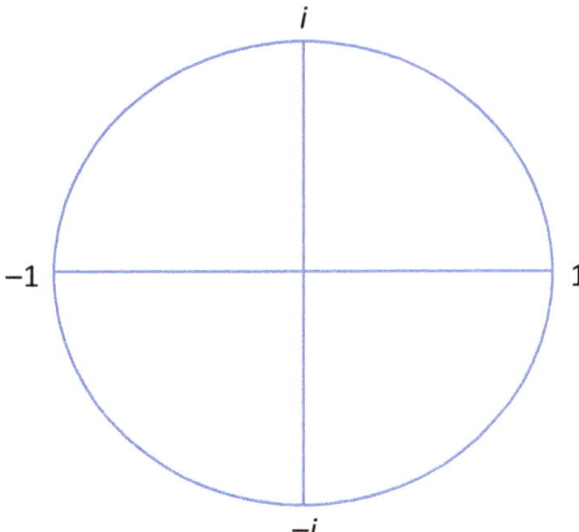

The payoff from creating mathematics leads through the centuries, to the development of modern physics where complex numbers are fundamental. Start with classical physics and a plane wave.

A plane wave is represented by

$$\psi = e^{i(kx - \omega t)}$$

which satisfies

$$\frac{\partial^2 \psi}{\partial t^2} = c^2 \frac{\partial^2 \psi}{\partial x^2}$$

with $c = \frac{k}{\omega}$.

If we change the variables to position and momentum and energy and time, we have now for a free particle with energy $\frac{p^2}{2m}$:

$$\psi = e^{i\left(\frac{px}{\hbar} - \frac{Et}{\hbar}\right)}$$

where $\hbar$ has the units to make the exponential argument dimensionless. The equation for ψ is

$$i\hbar \frac{\partial \psi}{\partial t} = -\frac{\hbar^2}{2m} \frac{\partial^2 \psi}{\partial x^2}$$

and if we add a potential term to the right-hand side $V(x,t)\psi(x,t)$, we get Schrödinger's equation for quantum mechanics. Apply this to a doped crystal and get semiconductors, apply this to atomic levels and get a laser. What started as an annoyance in solving quadratic equations when bumping into complex numbers has now become essential to understanding nature.

In ancient times, if a ruler asked for the payoff of playing with mathematics, it might have been hard to come up with a convincing answer. Nevertheless, mathematics advanced and blossomed. As just one example of the wisdom of leadership, in the

1700s, the St. Petersburg Academy supported Daniel Bernoulli and Leonhard Euler and outstanding work was achieved. Even today, when applying for grants, it may be hard to come up with a convincing argument for the payoff. Following the technological successes of World War II, this type of question, for science, was defended exquisitely in Vannevar Bush's 1945 book, *Science, The Endless Frontier.* Here was the defining declaration that set the program for funding basic science in the US. ONR was established in 1946, followed Bush's vision. Bush's arguments apply just as well to Mathematics: pure and applied. In fact, the last MURI project with which I was associated, at ONR, was the Mathematics of Deep Learning, an AI project. Basic research continues to create, discover, amaze, and change our world.

We can now have a comparative discussion of what was the rationale and motivation for developing physics. What was the payoff for physics? For the ancient Greeks and Romans, was it building better catapults or other simple machines? That was more engineering at which the Romans excelled, noting engineering and ingenious have the same Latin root. Did Archimedes design a lens to focus light to burn the sails of enemy ships? That could be a payoff. It is not clear to me why at the beginning theoretical physics was developed. What was the payoff? It seemed only to apply to mechanics. Electromagnetism was a way off and quantum physics beyond that. Maybe it was just there were people hardwired from birth striving to learn and understand, what today we call scientists. Apparently, Aristotle wrote that a constant force is needed to move an object at a steady pace, a reality with friction. Then, how does the Moon orbit the Earth? In Greek mythology, Apollo was the force needed to move the Sun across the sky. Isaac Newton was driven to understand beyond invoking a god. Even if he could write and solve an equation for an orbit, why would anyone care? Anyway, Newton has his famous force equation $F = ma$ where $F = \frac{dp}{dt}$ with a being the acceleration and p the momentum. Who knew where this would take math and physics. For $F = ma$ on a rotating frame of reference, in 1835, Coriolis derived his effect for the deflection of a trajectory to the right with application to long-range artillery targeting in the northern hemisphere. In 1861, for electromagnetism, Newton's start became Maxwell's equations:

$$\nabla \cdot E = \frac{\rho}{\epsilon_0}, \quad \nabla \times E = -\frac{\partial B}{\partial t}$$

$$\nabla \cdot B = 0, \quad \nabla \times B = \mu_0 J + \mu_0 \varepsilon_0 \frac{\partial E}{\partial t}$$

where $\mu_0 \varepsilon_0 = c^{-2}$ and getting back to force, there is in 1895 the Lorentz force due to electric and magnetic fields on a charged particle:

$$F = q(E + v \times B)$$

Modern electromagnetic technology has followed from these equations. Although these equations are linear, the interaction of light with matter is not.

Getting back to Newton, he not only understood inertia and invented calculus but also solved the nonlinear differential equation for the Earth–Moon or Earth–Sun system to derive a stable elliptic orbit and derive Kepler's laws. As the Newtonian equation derived a stable Moon orbit, much of the following work on differential equations was focused on stable solutions, such as the latter 1772 work by Lagrange on stable points that balance gravitational and centrifugal forces and are the home of satellites, man-made or natural, like the Trojan moons of Jupiter at these Lagrange points. But Newton had deeper thoughts and concerns. There was no friction in his equation and he worried that a phenomenon, such as Halley's comet, would disturb the orbit and there was no mechanism to smooth out such perturbations. So, why was the Moon's orbit stable? His attempt at a solution to a three-body problem Earth–Moon–Comet failed. In 1773, Laplace applied perturbation theory to prove, at least for a short time, the stability of the solar system, and in 1856, Maxwell proposed and calculated the stability of the rings of Saturn based on individual particles comprising the rings. The focus of science and engineering was on stability, not instability, but in 1893, Poincare in revisiting the question of stability of the solar system found the instabilities that today we call chaos. Too bad he didn't have a computer to visualize the complexity of chaos. It took until the 1950s to fully understand the dynamics of a (Hamiltonian) frictionless three-body system, including the possible very nonlinear behavior of chaos. An advance was the 1954 KAM theory for Hamiltonian systems showing that stable orbits could exist in a phase space chaotic sea. An example of this is Chirikov's standard map of a kicked rotator (mod 2π):

$$x_{n+1} = x_n + K\sin(\theta_n)$$

$$\theta_{n+1} = \theta_n + x_{n+1}$$

Here is shown a calculation by Gert Zumofen, Joseph Klafter, and myself, for $K = 1.1$ a stable period 7 orbit embedded in a chaotic sea. The colors reflect the time to exit from nearby points around the black stable orbit points.

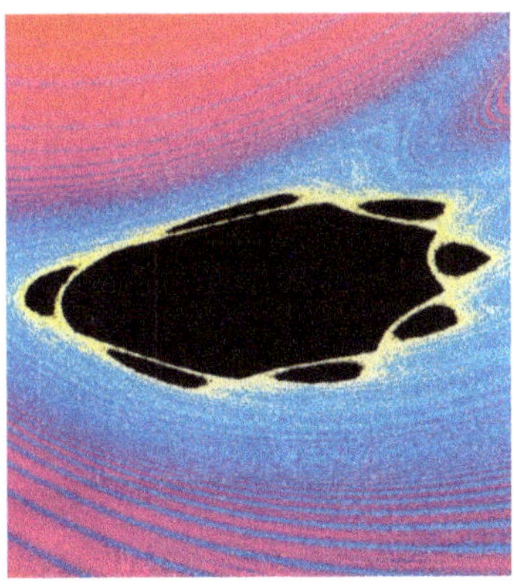

After KAM, the next major advance was due to Ed Lorenz, in 1963, studying fluid heat convection. He simplified the Saltzman 1962 partial differential equations to three coupled equations:

$$\frac{dx}{dt} = \sigma(y - x)$$

$$\frac{dy}{dt} = rx - y - xz$$

$$\frac{dz}{dt} = xy - bz$$

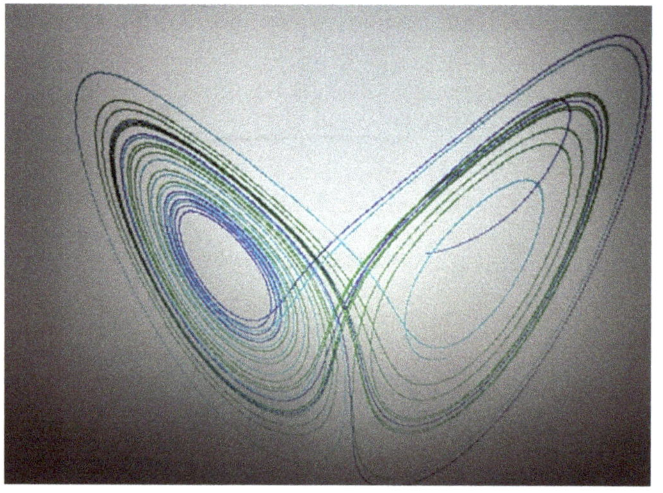

A Lorenz trajectory projected on the x–z plane with $\sigma = 11$, $r = 28$, $b = 8/3$. The trajectory starts on the right lobe and aperiodically spirals back and forth between the lobes. The colors change with time.

It took some time for Lorenz's work to be fully appreciated, but it is now regarded as revolutionary. One used to take a semester course in mechanics and then move on to E&M and quantum mechanics. But due to Lorenz and many others, now there exists a plethora of nonlinear dynamics topics to study.

I anticipated each year that the Nobel Prize would be awarded to Lorenz. I'm disappointed that he, with others, were never so honored. In 1985, I organized a conference at the Naval Surface Warfare Center, then at White Oak, Maryland. Ed Lorenz gave a great talk with ideas on how to analyze complex dynamics in higher dimensions and, in part, find the plane formed by the vector direction of the two largest Lyapunov exponents. Ed was the first one to submit his contribution "Atmospheric Models as Dynamical Systems" to our book, *Perspectives in Nonlinear Dynamics* (World Scientific Publishers, 1986).

Even today, in my ONR program, surprising advances can come from basic research. One example was the supporting work on chaos surprisingly leading to morphable logic gates. To the general public, math and science may be hard to grasp where progress occurs. Once, a US Senator known for looking for waste in science funding pointed to a study on the sex lives of insects. But the offshoot of the research was the fabrication of chemical odors that when sprayed on fields confused the insects to find mates and to die off without offspring, which led to protecting crops. The reader can probably find other examples.

While the benefits of the US federal government funding of science are appreciated, defending this funding is a major activity each year.

CHAPTER

A First Glimpse of Nonlinear Mysteries

Starting in 1970, I was a grad student in Elliott Montroll's group at the University of Rochester. Elliott was a renowned physicist holding one of NY State's Einstein Professorships. Being in Elliott's group, there was never any stress, just encouragement and joy in finding and working on good problems. One group member was Leon Glass, a postdoc. At group meetings, Leon discussed equations that were nonlinear with a rich set of properties. Even though I had been a math major at SUNY Stony Brook, I was ignorant about nonlinear equations and all that Leon was presenting. I was used to equations, for example, an oscillating LC circuit

$$\frac{d^2 q}{dt^2} = \frac{1}{\text{LC}} q\left(t\right)$$

whose solution oscillates with a frequency $\frac{1}{\sqrt{\text{LC}}}$. This is a linear differential equation, and nothing spectacular happens.

15

Look at the complicated-looking Bessel equation

$$x^2 \frac{d^2 J_0(x)}{dx^2} + x \frac{dJ_0(x)}{dx} + x^2 J_0(x) = 0$$

with solution

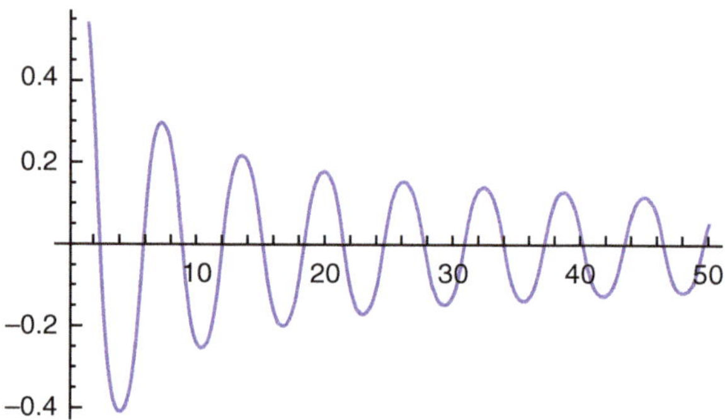

Nothing spectacular happens here either. There are many other linear examples where one gets a solution and that's it. It could be the partial differential equations for diffusion or the wave equation, and once you solve it, you're done. Leon gave examples where slightly changing an initial condition can cause a solution to asymptote to different values or even fly off to infinity. Other examples had solutions bifurcate from a single value to two or more values. Complicated behavior could arise by changing the value of a parameter. Every nonlinear equation seemed to have its own unique behavior, to be its own universe. Techniques, such as Fourier or Laplace transforms, matrix methods, and eigenvalue approaches, that worked so well with linear equations seemed not to be of value here. It did not seem, at least to me, that there was a theory of nonlinear equations, as there was for linear ones. Even iterating a quadratic map could produce complex behavior later called chaos. Going to nonlinear partial differential equations, a rich array of behaviors could be achieved by changing a single parameter, for example, the Reynolds number for fluid flow in the Navier–Stokes equation can change behavior from a laminar flow to a bubbly turbulent one. A variety of nonlinear equations could lead to dramatic changes in dynamics, including chaotic behaviors and associated fractal geometry. There are today, fresh new topics for the modern math and physics curriculum. Little did I know in 1970 that nonlinear equations would be my focus for 40 years at ONR starting in 1983.

The Richness of Nonlinear Equations: A Few Basic Examples

An example of a nonlinear iterative map is Newton's method for finding the roots of an equation $f(x) = 0$. We will look at $f(x) = x^3 - 1$ with roots at $x = 1$, $x = e^{\frac{2\pi i}{3}}$, $e^{\frac{4\pi i}{3}}$. Using Newton's iteration formula for the nth iterate

$$x_{n+1} = x_n - \frac{f(x_n)}{f'(x_n)}$$

if one starts the process near one of the roots, the solution will converge toward that root. However, there are interlaced regions where a small change in the initial guess x_0 will change to which root the solutions will converge. The basin boundaries for the roots are color-coded and present a rich geometry. The cubic root basin boundaries are used for this book cover. Note that the region colored green contains the initial starting points that converge on the root $x = 1$. For the equation

$$f(x) = x^5 - 1$$

the fifth roots of unity basin boundaries are shown with five colors:

The basin boundaries for the third roots and fifth roots of unity.

When I started programming in 1968, it was either in FORTRAN on the IBM 360 or a basic language with the PDP-9. I calculated numbers, but no visual display. It was a pleasure to much later learn (the now ancient) TrueBasic in the 1980s where I could plot visual information, such as the above cube roots and fifth roots of unity. The photos of the cube and fifth roots' domains of attraction are from my old laptop screen; otherwise, its output would be on a floppy disk, and I no longer have a floppy disc reader.

Another elementary example of a nonlinear equation is

$$\frac{dx(t)}{dt} = x^2 - a^2$$

Let's set $a = 1$, then plotting the derivative versus x, we see that it is zero at ± 1:

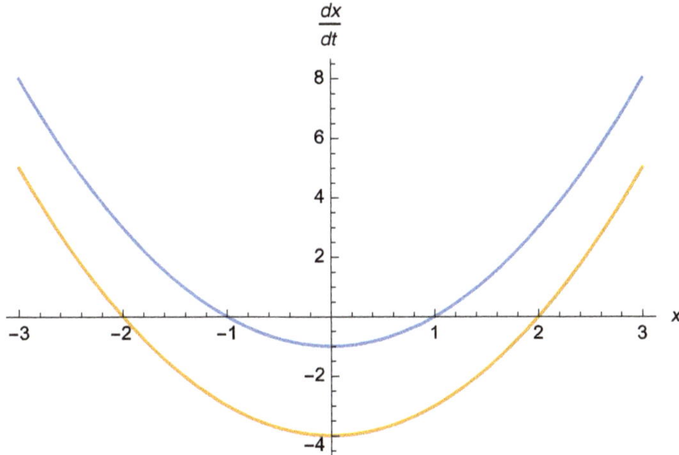

Note that $x = -1$ is an attracting point as the derivative dx/dt is positive to the left of -1, so in time, the solution moves to the right, while for x in $(-1, 1)$, the derivative is negative and the solution moves to the left. Any initial starting point for x in $(-\infty, 1)$ the solution will asymptote to $x = -1$. For an initial condition greater than 1, dx/dt is positive, so the equation's time evolution goes to infinity. For x values less than 1, the solution moves to $x = -1$. Moving toward a stable point is a type of habituation. Start near $x = -1$, add a little noise, and the solution comes back to $x = -1$, basically ignoring the noise. If I now chose a different "a" value, say $a = 2$, the position of the stable point moves to $x = -2$ and the repelling behavior to infinity will change with $x = 2$ replacing the behavior formerly at $x = 1$.

Years later, when I was working at the Office of Naval Research, a submarine captain told me that when executing a fast turn, a submarine might fall or rise in depth. I said there must be a cubic nonlinearity in the water flow around the sub's sail.

Consider the nonlinear equation where z is the depth:

$$\frac{dz(t)}{dt} = az - z^3$$

Choosing $a = -1$, then $\frac{dz(t)}{dt}$ is positive for negative values of z, moving $z(t)$ toward zero, and $\frac{dz(t)}{dt}$ is negative for positive values of z, moving $z(t)$ to the left. Thus, for $a \leq 0$, the value $z = 0$ is a stable fixed point.

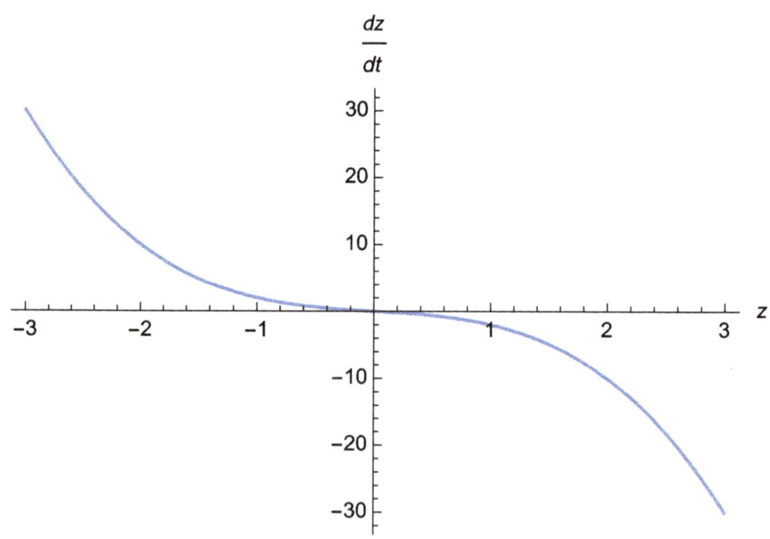

For $a > 0$, the origin becomes unstable and two stable points arise as shown in the following. Choose $a = 4$, then 2 and -2 are the new stable points; in my discussion, these represented the submarine increasing or decreasing depth by 2. This is called a pitchfork bifurcation, as one stable solution bifurcated into two prongs. When I presented this little calculation to the Navy lab at Carderock, Maryland, they showed no interest and instead were working on a neural network approach.

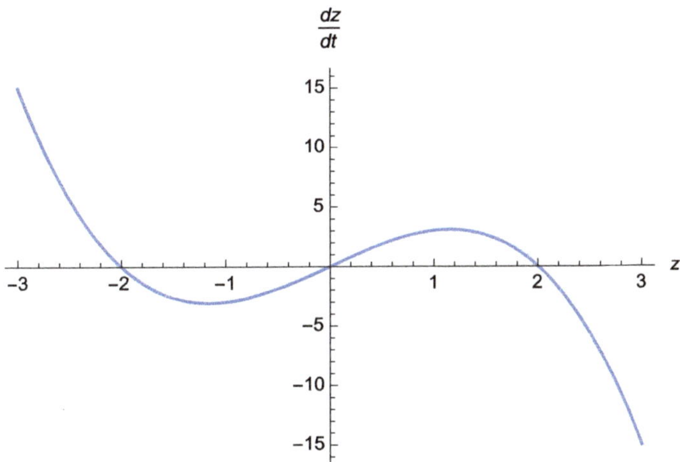

In any event, Leon Glass' talks made nonlinear equations seem that each equation was its own universe with unique behaviors, and small changes could result in a variety of outcomes, nothing like linear equations, and I did not pursue any research in that direction.

In contrast to linear equations, nonlinear equations seem alive. Behaviors can change with a parameter change. Consider this quadratic example:

$$\frac{dn(t)}{dt} = -E(k)n(t) - Bn^2(t)$$

When $E(k) > 0$, the right-hand side is always negative for any positive n and $\frac{dn(t)}{dt}$ is negative (the lower curve in the following), making the origin the only stable point.

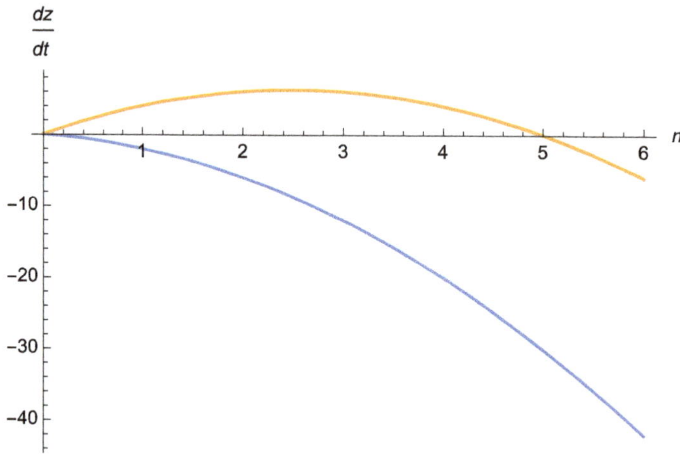

For an $E(k) < 0$ (the upper curve), the origin now becomes a repeller and $n = \frac{-E(k)}{B}$ becomes a stable point.

If k oscillates making $E(k)$ oscillate between positive and negative values, then the solution switches between $n = 0$ and $n = \frac{-E(k)}{B}$, which is reminiscent of slip–stick frictional motion.

These have just been a few simple examples of nonlinear behaviors of creating complex basin boundaries, creating and moving stable and repeller points. A great place to learn about these types of topics and more is *Nonlinear Dynamics and Chaos* by S. Strogatz. A very friendly readable text.

Some differences between linear and nonlinear equations are as follows:

Linear	vs	*Nonlinear*
Waves		Solitons
Diffusion		Chaos
....		Intermittency
Superposition		Mode competition
Completeness		Reduction of degrees of freedom
Integrability		Strange attractors
Hamiltonian stable orbits		Strange kinetics
Euclidean geometry		Fractal geometry
Smoothness		Bifurcations

Some other interesting things were around in 1970, of which I was unaware, such as the 1954 KAM theorem of stable orbits in a chaotic sea for conservative systems and Lorenz's chaotic butterfly attractor in 1963. There was also Gollub and Swinney's 1975 experimental quasiperiodic route to chaos in a fluid when a third incommensurate frequency is generated, adding to the two existing ones. This differed from Landau's theory that expected the need to generate an infinite number of frequencies to reach what we now call chaos. Years later, the topic of chaos exploded with interest when Feigenbaum discovered a universality in the period-doubling route to chaos. More about that later and the story of how I started my 40-year career at ONR, but now something about linear equations and eigenvectors and eigenvalues.

In Praise of Linear Equations: Designing Peptides

In this chapter, I will specifically focus on the topics of eigenvectors and eigenvalues of coupled linear equations. The objective will be to use the components of the eigenvectors to design peptides. The mathematical question is to consider a string of N numbers and then to find a smaller string of numbers m that has the same spectral properties. N will correspond to the number of amino acids in a transmembrane receptor protein. This is a class of proteins that form linear chains that go into and out of a cell membrane seven times and are called G7 transmembrane receptor proteins. These various receptors control brain chemistry by binding neurotransmitters like dopamine and serotonin.

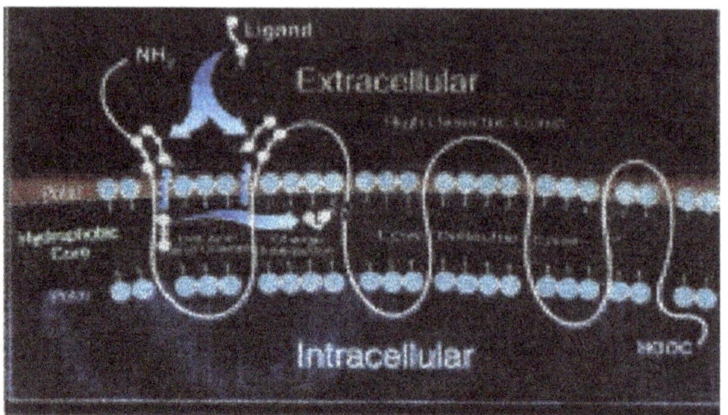

Each amino acid (there are 20 that are used to build proteins) has a measured affinity to attract (hydrophilic) or repel (hydrophobic) water.

Tanford Scale of Amino Acid Hydrophobicity

Name	Symbol	kcal/mol	Name	Symbol	kcal/mol
Glycine	G	0.0	Cysteine	C	1.52
Glutamine	Q	0.0	Lysine	K	1.65
Serine	S	0.07	Methionine	M	1.67
Threonine	T	0.07	Valine	V	1.87
Asparagine	N	0.09	Leucine	L	2.17
Aspartate	D	0.66	Tyrosine	Y	2.76
Glutamate	E	0.67	Proline	P	2.77
Arginine	R	0.85	Phenylaline	F	2.87
Alanine	A	0.87	Isoleucine	I	3.15
Histidine	H	0.87	Tryptophan	W	3.77

With N about 500, the G7 receptor protein tends to have the hydrophobic amino acids within the cell membrane and the hydrophilic ones inside the cell or outside in the cellular fluid.

We started by treating the amino acid sequence as a time series $s(t)$ and performing a wavelet transform $g(t)$.

We chose the complex Morlet wavelet, $T(a, b)$

$$T(a, b) = \frac{1}{\sqrt{b}} \int_{-\infty}^{\infty} s(t) g\left(\frac{t - a}{b}\right) dt$$

with the complex wavelet

$$g(t) = e^{-\frac{t^2}{2}} e^{2\pi i \omega t}$$

to produce 2D plots for the real and imaginary $T(a, b)$. The imaginary plot for Hemoglobin β, which is dominated by alpha helices, is shown in the following:

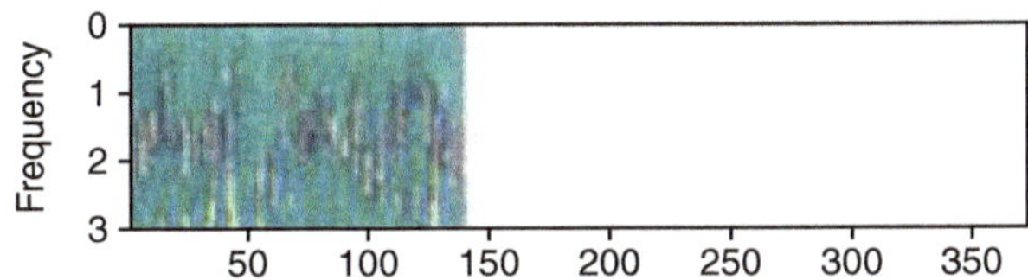

This is contrasted with concanavalin A, which is dominated by β sheets,

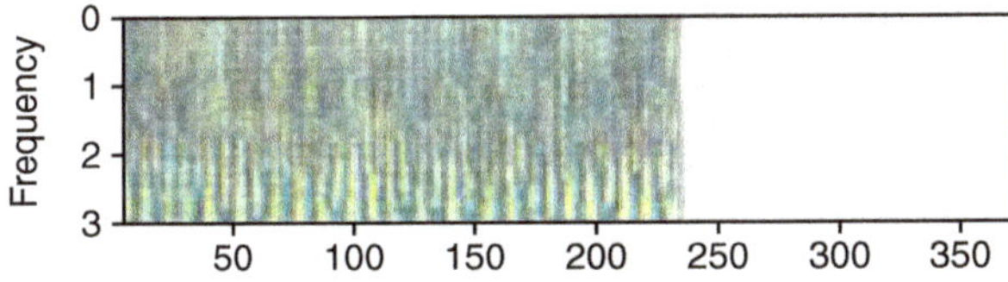

demonstrating that this method differentiated between dominant helix or sheet structures from the amino acid sequence. These plots were due to Norden Huang, from NASA Goddard, who did prior wavelet calculations for the studies of ocean waves. I gave Norden the amino acid sequence hydrophobic values, to which he applied his ocean wave analysis approach. This was the benefit of being in the Washington DC area with so many opportunities for expert help. Norden is famous for the Hilbert–Huang transform to treat a signal in terms of an empirical mode decomposition.

This work was a joint effort with Arnold Mandell and Karen Selz from the Cielo Institute. More about Arnold later, as he influenced me to create a national nonlinear science program at ONR. Our next step was to use linear algebra to design small peptides to bind to specific G7 transmembrane proteins. We label each amino acid by its measured hydrophobicity value, so that we have a sequence h_1 to h_N for an N-length G7 protein. The objective is to capture the spectral properties of the N-number sequence in a shorter m-number sequence. We choose $m = 15$, as that is the size of the peptide protein that we want to design to bind to the G7 transmembrane protein and have the peptide be small enough to pass through the blood–brain barrier. We start by constructing the following $15 \times (N - 15)$ matrix:

$$\begin{pmatrix} h_1 & \cdots & h_{N-14} \\ h_{15} & \cdots & h_N \end{pmatrix}$$

Let $N - 15$ be denoted as A. This is a $15 \times A$ matrix and we multiply it by its adjoint $A \times 15$ to generate an $M = (15 \times A)(A \times 15) = 15 \times 15$ matrix. Let $K = N - m + 1$, choosing $m = 15$, then we have the following:

$$M = \begin{pmatrix} \sum_{i=1}^{K} h_i h_i & \cdots & \sum_{i=1}^{K} h_i h_{i+m-1} \\ \sum_{i=1}^{K} h_i h_{i+m=1} & \cdots & \sum_{i=1}^{K} h_{i+m-1} h_{i+m-1} \end{pmatrix}$$

This matrix has 15 eigenvalues, each associated with a 15-component eigenvector. We call these Broomhead–King modes.

We pick an eigenvector (we choose the one with the second largest eigenvalue) and plot the 15 eigenvector components as a series. Now, from the 20 amino acids, we choose a sequence whose h values best match the eigenvector components series in shape. We then explore if this $m = 15$ sequence captures the essence of the large N around the 500 number sequence. In fact, experiments were performed that showed several of our peptides designed to bind to dopamine receptors could change the pH of a cell and in experiments with rats could change their behavior in a maze.

Using the amino acid symbols from the chart, we found that SHQRWEYKGVNCIVY had a direct significant effect on the cell acidification of the second dopamine receptor in mouse fibroblast kidney cells.

A method to tell if the short m sequence captured the spectral features of the long N sequence is to compare the power spectra. The standard power spectrum for a sequence of data points $x(k)$ is an all-zeros polynomial

$$S(\omega) = \left| \sum_k x(k) e^{ik\omega} \right|^2 = \sum_{k=-\infty}^{\infty} c_k e^{i\omega k}$$

This will need many terms to capture peak behaviors. For a short sequence, one can try the all-pole power spectrum

$$S(\omega) = \frac{1}{\left| 1 + \sum_{k=1}^{m} a_k e^{i\omega k} \right|^2}$$

and expand in powers of $i\omega$ to match the $c_k e^{i\omega k}$ powers. The near zeros of the denominator produce peaks in the all-poles spectrum that would need many terms in the all-zeros power spectrum to produce. For example, compare the power spectrum for the many-unit estrogen receptor alpha with our all-pole power spectrum for the designed peptide YGGVAREWFFLISKE:

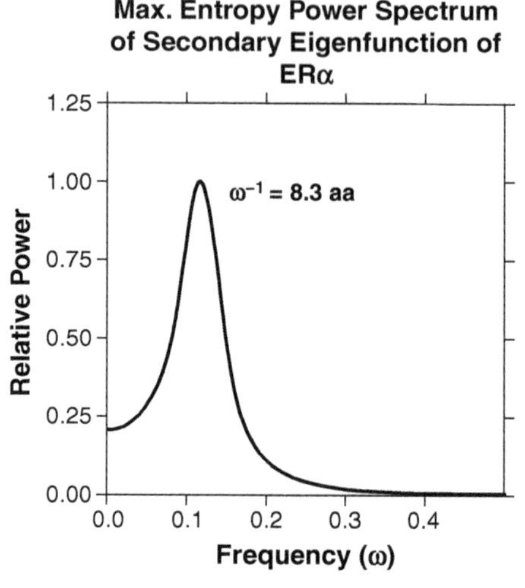

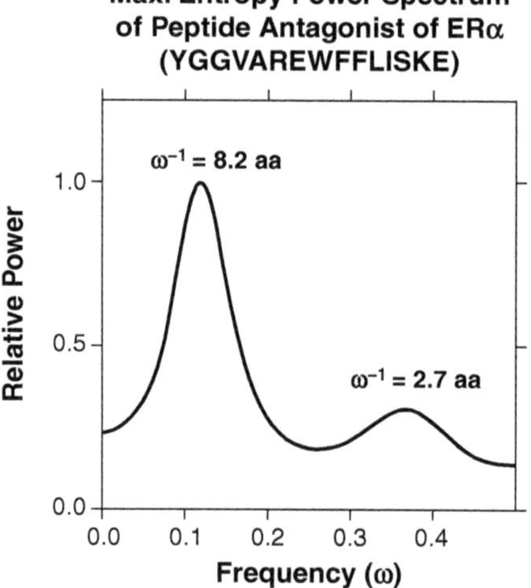

Arnold, Karen and myself have patented this work:

US007228238B2

<table>
<tr><td>

(12) **United States Patent**
Mandell et al.

</td><td>

(10) **Patent No.:**　**US 7,228,238 B2**
(45) **Date of Patent:**　*Jun. 5, 2007

</td></tr>
</table>

(54) **ALGORITHMIC DESIGN OF PEPTIDES FOR BINDING AND/OR MODULATION OF THE FUNCTIONS OF RECEPTORS AND/OR OTHER PROTEINS**

(75) Inventors: **Arnold J. Mandell**, Asheville, NC (US); **Karen A. Selz**, Asheville, NC (US); **Michael F. Shlesinger**, Rockville, MD (US)

(73) Assignee: **Cielo Institute, Inc.**, Asheville, NC (US)

(*) Notice: Subject to any disclaimer, the term of this patent is extended or adjusted under 35 U.S.C. 154(b) by 455 days.

This patent is subject to a terminal disclaimer.

(21) Appl. No.: **10/777,829**

(22) Filed: **Feb. 11, 2004**

(65) **Prior Publication Data**

US 2005/0027457 A1　　Feb. 3, 2005

Related U.S. Application Data

(63) Continuation of application No. 09/767,460, filed on Jan. 23, 2001, now Pat. No. 6,865,492, which is a continuation-in-part of application No. 09/490,702, filed on Jan. 24, 2000, now Pat. No. 6,560,542.

(51) **Int. Cl.**
　G06F 19/00　　　(2006.01)
　C07K 2/00　　　(2006.01)
(52) **U.S. Cl.** **702/19**; 530/300
(58) **Field of Classification Search** .... None
See application file for complete search history.

(56) **References Cited**

OTHER PUBLICATIONS

Mandell, A.J. (1984) Non-equilibrium behavior of some brain enzyme and receptor systems. Ann. Rev. Pharm. Toxicol. 24:237-274.
Mandell, A.J., Selz, K.A. and Shlesinger, M.F. (1997) Mode matches and their locations in the hydrophobic free energy sequences of peptide ligands and their receptor eigenfunctionc. Proc. Natl. Acad. Sci. 94:13576-13581.
Mandell, A.J., Selz, K.A. and Shlesinger, M.F. (1997) Wavelet transformation of protein hydrophobicity sequences suggests their memberships in structural families. Physica A224: 254-262.
Mandell, A.J., Selz, K.A. and Shlesinger, M.F. (1997) Hydrophobic free energy eigenfunctions help define continuous wavelet trans-

Di Marzo, E.A. and Mandell, A.J. (1997) Phase transition behavior of a linear macromolecule threading a membrane. J. Chem. Physics 197:5510-5514.
Mandell, A.J., Owens, M.J. Selz, K.A., Morgan, W.N., Schlesinger, M.F. and Nemeroff, C.G. (1998) Mode matches in hydrophobic free energy eigenfunctions predict protein—protein interactions. Biopolymers 46:89-101.
Selz, K.A., Mandell, A.J., and Shlesinger, M.F. (1998) Hydrophobic free energy eigenfunctions of pore, channel and transporter proteins contain B-burst patterns. Biophysical J. 7:2332-2342.
Mandell, A.J., Selz, K.A., Shlesinger, M.F., and Kuhar, M.J. (1999) Linear and entropic transformations of the hydrophobic free energy sequence help characterize a novel brain polyprotein: CART. In (M.T. Batchelor and L. Wille, eds.), *Statistical Physics on the Eve of the Twenty-First Century*. World Scientific, NJ, pp. 131-152.
Doyle, P.M. (1995) Combinatorial Chemistry in the Discovery and Development of Drugs. J. Chem. Tech. Biotechnol. 64:317-324.
Gordon, E.M., Barrett, R.W., Dower, W.J., Fodor, S.P.A. and Gallop, M.A. (1994) Applications of Combinatorial Technologies to Drug Discovery. 2. Combinatorial Organic Synthesis, Library Screening Strategies, and Future Directions. J. Med. Chem. 37(10):1385-1401.
Houghton, R.A. (1993) The Broad Utility of Soluble Peptide Libraries for Drug Discovery. Gene 137:7-11.
Mandell, A.J., Russo, P.V. and Blomgren, B.W. (1987) Complex hydrophobic sequence transformation predicts mutual recognition by polypeptides and proteins. Ann. N.Y. Acad. Sci. 504:88-118.
Mandell, A.J., Selz, K.A. and Shlesinger, M.F. (1998) Transformational homologies in amino acid sequences suggest membership in protein families. J. Stat. Phys. 93:673-697.
White et al. (1990) Statistical distribution of hydrophobic residues along the length of protein chains, Biophys. J., vol. 57 pp. 911-921.
White, Stephen H. (1994) Global Statistics of Protein Sequences: Implications for the Origin, Evolution, and Prediction of Structure. Annu. Rev. Biophys. Biomol. Struct. 23:407-439.
Chorev M. et al. "Recent Developments in Retro Peptides and Proteins—An Ongoing Topochemical Exploration", Trends in Biotechnology, Elsevier, Amsterdam, NL., vol. 13, No. 10, Oct. 1995, pp. 438-445, XP004207219 ISSN: 0167-7799.
Raffa: "Drug-Receptor Thermodynamics: Introduction and Applications," May 2001, John Wiley & Sons XP001153602. Mandell et al: Hydrophobic Mode-Targeted, Algorithmically Designed Peptide Ligands Structure as Modulators of Protein Thermodynamic Structure and Function p. 655-p. 700.

Primary Examiner—John S. Brusca
(74) *Attorney, Agent, or Firm*—Wilson, Sonsini, Goodrich & Rosati

(57)　　　**ABSTRACT**

Methods of synthesizing a peptide or peptide-like molecule to a polypeptide or protein target based on mode-matching each member of a set of peptide constituents of the peptide or peptide-like molecule to peptide constituents of the target polypeptide or protein target.

Note that one can play with wavelets just for fun using Mathematica. Here, I did one, treating a series of 27 zeros and ones as a sequence to create the flavor and colors of a Jamaican flag:

The Surprise in Quadratic Maps

The equation $\frac{dx}{dt} = rx(1 - x)$ known as the logistics equation is readily solved as $x(t) = \frac{1}{1+e^{-rt}}$.

That's the type of math I learned in my differential equations course. Nothing spectacular happens; start with $x(t = 0) = 1/2$, and x grows to 1 as time approaches infinity. Now, go to the discrete version,

$$x_{n+1} = \lambda x_n(1 - x_n)$$

and a whole new world opens up that was not mentioned in my courses. Iterate the discrete equation and plot for $\lambda \leq 4$ to find an increasingly complicated behavior as λ increases, involving period doubling bifurcations. Note that for $\lambda > 4$, the trajectory diverges. This divergence is an example of a dynamical "crisis" where the attractor hits a repelling point at its boundary and the trajectory goes to infinity. Celso Grebogi, Edward Ott, and

James Yorke were a powerhouse of new results for nonlinear dynamical systems, including new results for types of crises. See Ott's book, *Chaos in Dynamical Systems*, for an excellent exposition of many aspects of chaos. Following are some plots from my ancient TrueBasic programming.

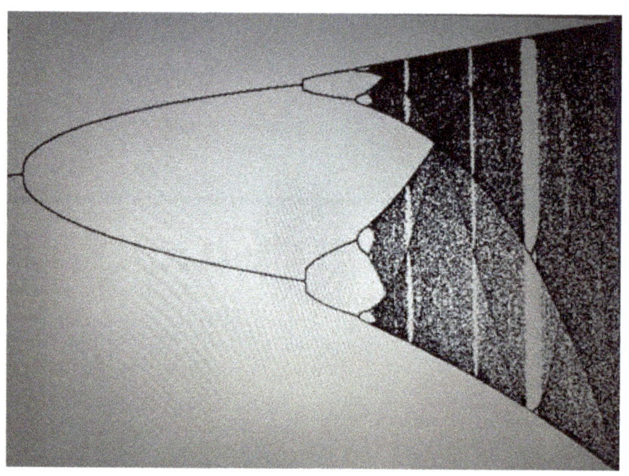

The famous plot of x versus λ up to $\lambda = 4$.

Next, here are some examples of the behavior for some λ values, plotting iterates of x_{n+1} vs x_n for a few n values:

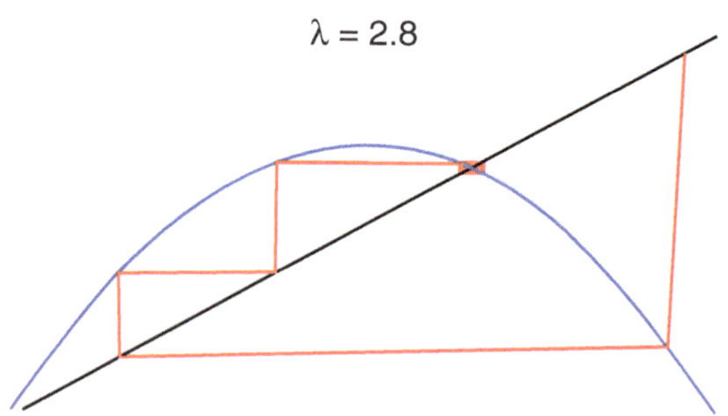

At $\lambda = 2.8$, the iteration converges rapidly to a single value.

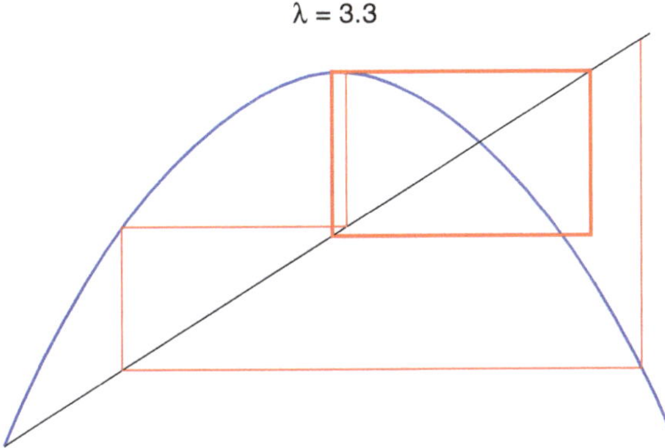

Here, at $\lambda = 3.3$ the solution bifurcates to two values, and for $\lambda = 3.5$, there are further bifurcations:

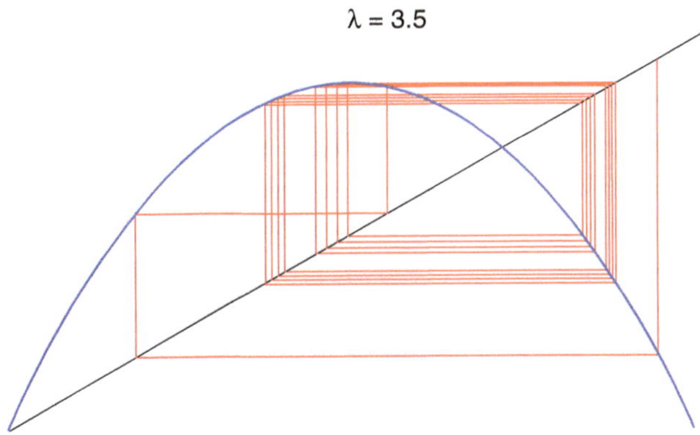

First, there is the surprise that something viewed as a map could be so different than the continuum version. Second, it is a surprise that something seemingly as simple as a quadratic map could produce so complex a result when a parameter is varied.

It becomes even more interesting when we generalize from real values x to complex values z. In fact, one can choose the variable to be real, complex, quaternion, or even octonian.

S. Grossman and S. Thomae for real x found for the λ_n values for the nth period doubling bifurcation that

$$\lim_{n\to\infty} \frac{\lambda_n - \lambda_{n-1}}{\lambda_{n+1} - \lambda_n} = 4.669\ldots$$

Well, $4.669\ldots$ is just a number. The excitement arose when Mitchell Feigenbaum found the same number for other equations that exhibited period doubling. For example, Feigenbaum demonstrated that the equation

$$x_{n+1} = \lambda\sin(x_n)$$

had the same properties as the quadratic map, finding again the 4.669 value, demonstrating only the shape of the right-hand-side mattered (a unimodal peak) and not the specific function. Here's the sine plot.

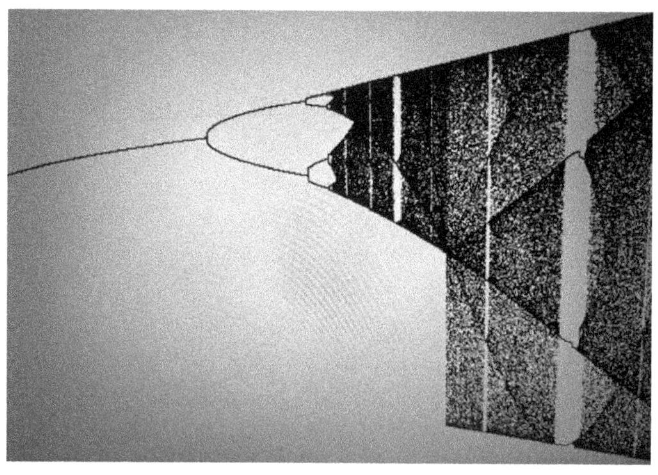

Imagine, before computers, how difficult it would be to generate this plot. Years ago, again with ancient TrueBasic, this is all it took for me to program the above sine plot:

```
SET WINDOW 1, 4, −4.5, 4.5
LET X = PI/7
FOR XLAM = 1 TO 4 STEP .0001
FOR XTIME = 1 TO 10 STEP 1
LET X = XLAM*SIN(X)
PLOT XLAM, X
NEXT XTIME
NEXT XLAM
END
```

Feigenbaum's work led me to want to study nonlinear dynamics. It would be hard to win a grant, to support myself, on a topic in which I had produced no work, but I could learn the field and be immersed in nonlinear physics as a Scientific Officer at ONR. From one topic and paper to the next, I never knew what problem I would work on. You hear a seminar, read a journal article, talk to a friend, and that gets you started on a new unexpected topic. That's why I told people that I funded through ONR, that I do not expect them to know what they will be working on through the grant term.

As there are many excellent texts on nonlinear dynamics, I am not trying to duplicate their work and detailed mathematical analyses but only to provide the influences on my personal trajectory in physics.

The math was nice, but was any of this relevant? Albert Libchaber, in 1980, found period doubling with the Feigenbaum constant 4.669 for heated convection experiments in mercury.

After this, nonlinear dynamics became viral. Period doubling and chaos were discovered in electron–hole oscillations in Ge, in 1984, by Carson Jeffries' group and studied by Robert Westervelt's group at Harvard. Chaos was found in a variety of physical systems when driven hard enough into nonlinear responses. Was chaos found earlier? Ralf Landauer from IBM pointed out from his time in the US Navy, as an Electrician's Mate, that there was a US Navy electronic technician training program in 1945 with an exercise to build seemingly random blinking lights, what we would call chaotic behavior and was called an idiot box. It was a battery-driven coupled circuit that someone back then appreciated that coupled oscillators could have an interesting irregular behavior. After theory and experiments, the next stage is applications and devices. I'll return to applications with a discussion of projects at Navy labs.

On the mathematics front, the study of quadratic mappings in the complex plane had a long history. Gaston Julia, in 1918, considered the quadratic equation for points in the complex plane (unfortunately, the computer was in the future to produce complex drawings):

$$z_{n+1} = z_n^2 + c$$

For c fixed, the Julia set is the boundary of the points that do not approach infinity upon iteration. The Mandelbrot set is the set of c values upon iteration the z values remain finite. These sets lead to fractal patterns of exquisite beauty with unending detail upon magnification. See the *Beauty of Fractals* by H.-O. Peitgen and P. H. Richter (Springer-Verlag, 1986).

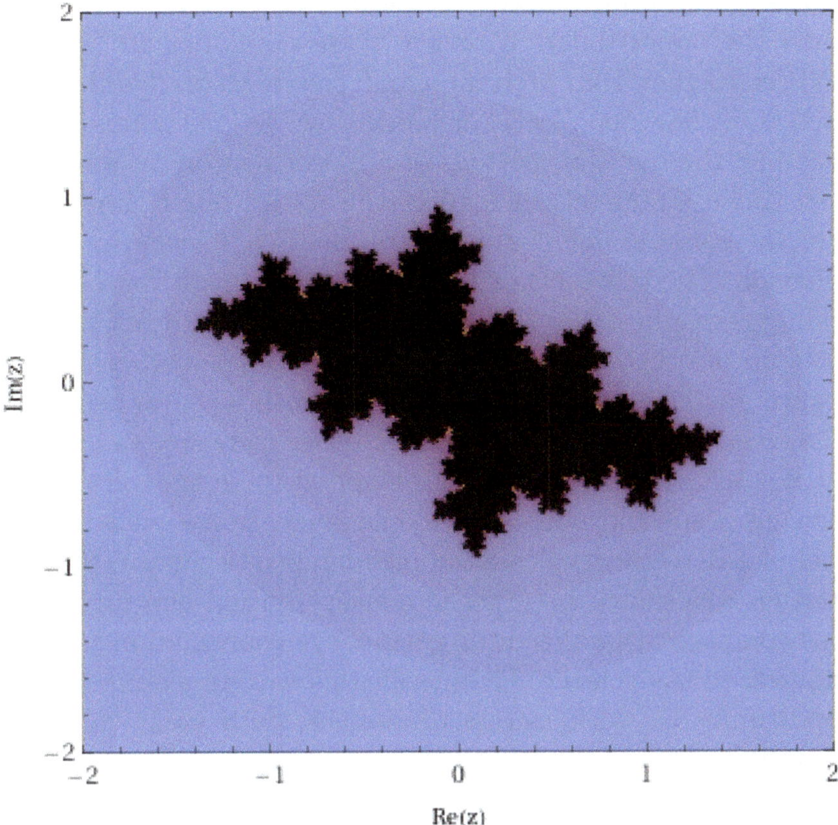

Filled in Julia set with $c = -0.5 + 0.5i$ that I computed with Mathematica.

Could the quadratic map play a role in biology? The mathematics of nonlinear dynamics has now been providing the tools to contribute to the analysis of complex biological systems that are nonlinear, non-equilibrium, non-stationary, open, and strongly coupled. This has been especially true for neuroscience, from single neuron bursting patterns to global EEG measures. I had the opportunity to review these developments with my ONR colleagues Tom McKenna and Teresa McMullen in Neuroscience *60*, 587–605 (1994) on work that we jointly funded. For example, the bursting pattern of a motor neuron, in a mollusk, fit a single hump return map of the $n{+}1$st burst time vs the nth, the same as for the quadratic map that leads to chaotic behavior. There are other examples of nonlinear mathematics finding a role in biology. Here, we mention two topics that will show up in other contexts in work at Navy labs: stochastic resonance and chaos synchronization. Crayfish receptor sensitivity followed a single humped curve versus temperature as an example of stochastic resonance with temperature in the role of noise, showing a specific noise level could increase a signal to noise level of a detector, typically an overdamped oscillator. Synchronization was shown between a monkey's visual and motor neurons complex dynamics during hand–eye coordination tasks. In addition to the science, for myself, this was an example of the opportunity to work, across disciplines, with great colleagues at ONR.

IPST Past and Future

At the University of Rochester, I was a graduate student in Elliott Montroll's group from 1970 to 1975 and then a post-doc until 1977. I left Montroll's group in 1977 to join Georgia Tech as a Research Scientist. When Montroll retired from the U of Rochester, he moved to the University of Maryland in 1980, rejoining his old position at the Institute for Physical Science and Technology (IPST). He asked me to join him there. In 1983, there was an external 5-year review of IPST headed by John Deutch from MIT. I knew John through Montroll, as well as knowing John's thesis advisor, Irwin Oppenheim, also at MIT in a physical chemistry group. Also in the MIT group was Bob Silbey as the Department Chair. My colleague/co-author and friend Yossi Klafter postdoc'ed with Bob. Yossi eventually became the President of Tel-Aviv University. John eventually was a Deputy Secretary of Defense and later Director of the CIA. Even though I was essentially at IPST as Montroll's postdoc, the review panel recommended me to be the new Head of the Institute. Montroll and the present Head, Joe Silverman, backed my nomination. Bob Zwanzig, as the leading theorist, favored Bob Dorfman, a physicist and long-time member of the Institute. Dorfman, by far, was the wisest choice, and he assumed the job. After a year, the U of Maryland Dean of Science left, and Bob then assumed that role. Then, he was promoted to Provost and then to Vice President of the university. As Vice President, in 1986, Bob offered me to return, from ONR, to the university to head a new Applied Math department. The pure and applied mathematicians were looking to split, with an applied math department open for internal sabbaticals to all members of the university. It would welcome economists, biologists, etc. When the question came to the Board of Regents, the creation of an Applied Math Department was shot down. The reasoning was that it would leave the Math Department too weak, as the pure mathematicians rarely published and were not well supported by grants. Then, Bob offered me a half-time position at IPST (where teaching was not required) and half-time in the Math Department. I gave a standing room only seminar on fractals (then a new topic) at the Math Department, but the pure mathematicians didn't attend. I was asked to give a second seminar for them but declined and said to only contact me again with an offer, which never came. Nevertheless, a happy ending. I stayed at ONR and was promoted to the Senior Executive Service in 1987, and in 2004, IPST honored me with their Distinguished Postdoc Alum Award for my time 1980–1983.

By the way, Bob Dorfman eventually left the Vice President position and returned to IPST. Although higher administration positions may sound prestigious, being a researcher is more fun. See Bob's book, *An Introduction to Chaos in Nonequilibrium Statistical Mechanics*. Scientific works are remembered, and administrators can be forgotten the day after they are replaced. In a search for a Dean of Science at the University of Maryland, the theme was to clean up the financial mess left by the last Dean. Maybe many searches are like that. Maybe Presidential races are posed as that.

CHAPTER 9

The Path to the Office of Naval Research

The first paper I published in 1973 (with V. M. Kenkre and E. W. Montroll), *Generalized master equations for continuous-time random walks, J. Stat. Phys 9*, 45 (73), acknowledged an ARPA grant managed by ONR. The grantee was Elliott Montroll, Nitant Kenkre was a postdoc, and I was a graduate student. I had no concept of what ARPA or ONR were. Note that ARPA latter became DARPA. I'll say more about DARPA later. However, I doubt that Montroll applied for a grant on this topic. He had a sterling reputation, with several past DOD connections, and I believe in those early days that was sufficient to get funding. He used to alternate the acknowledgments on published papers between the several grants that he had. This one was to ONR; maybe the next one would acknowledge ARO or NSF. Years later, as a Scientific Officer at ONR, my program had a focus, predicting, exploiting, or avoiding nonlinear dynamical behaviors. Well-known people would call up to apply for a grant, and I would explain that their proposed work is outside the interest of my program. Inevitably, they would say in the old days, good people got funded regardless of the proposed topic. As funding was limited, even good work in nonlinear physics that I would have liked to fund had to be passed over. As I needed to annually defend my program, it could happen that a good proposal covered topics that were already in my program, so no funding was forthcoming. Still, there are several government funding programs for physicists to try, including ONR, ARO, AFOSR, and DARPA, just within the DOD; others include NSF, DOE, NIH, NASA, and private foundations.

Getting back to my time at the University of Maryland, to get a grant, Montroll and I went together to the NSF, then in DC at 18th and G street. Being a distinguished professor, I thought Elliott would have a taxi bring us to the NSF. But Elliott was very down to earth. So, from the university, we walked to route 1 and caught a bus to the Rhode Island Metro stop for the Metro trip downtown. It was my first time on the Metro, and what a difference from the New York City subway. It was clean, uncrowded, and quiet. Arriving at the Farragut North metro station, we walked to the NSF with a stop for donuts close to NSF. Elliott was applying for a university/industry cooperation grant. The industry partner would be Harvey Scher, our colleague and friend at Xerox with whom we had published works. Harvey had now moved to SOHIO. (Later, when BP bought SOHIO, Harvey moved to the Weizmann Institute in Israel. He switched from his earlier work on electrons in disordered semiconductors to chemical dispersion through water tables). The talk with the NSF Program Manager was going well and funding seemed assured, but then another Program Manager entered and addressed Elliott asking if he was Dr. Montroll. He asked about the final report. What? At the University of Rochester, Elliott had a large multi-year NSF grant to fund his Institute of Fundamental Studies. When I joined, I was Elliott's 10th graduate student, and there were several postdocs and affiliated faculty. When the grant ended, there needed to be a final report, including what papers were published, who was paid, who graduated and when, etc. Elliott planned for one of the Assistant Professors to handle this, but that person left for a different job opportunity. The final report was never written, and the grant was never officially completed and closed. The NSF manager said no new grants until the last one was closed out. We left with no grant but a new idea.

Elliott had been recently a member of JASON, a high-level, academic-based DOD advisory group. The group was funded through DARPA. Elliott recently retired from the University of Rochester and, when turning 65, also retired from JASON. We next went to DARPA, then in Rosslyn, Virginia, to seek a grant from the DARPA manager who oversaw the JASON funding. I think it was in April 1981. The government fiscal year starts on October 1st. The DARPA manager said, it's late in the fiscal year and most funds are already committed, the best I can do is $250K. Wow, that was a large sum in 1981 for basic research. For advanced research projects, DARPA could easily deal in the millions. The grant proposal was submitted through Montroll's company, the La Jolla Institute (LJI). So, Bruce West (a former Montroll postdoc), a leader at LJI, and I wrote up a proposal, and we got the DARPA funds. I was both at Maryland and LJI and was paid with the DARPA funds through the University of Maryland. My Maryland salary, in 1980, when I joined IPST, was $22K. Now, listed as a senior LJI scientist, my salary was $38K, which gave me enough confidence to buy a townhouse, located about halfway between NIH and NBS (now named as NIST).

LJI had an earlier grant from ONR in the area of fluid mechanics. The story that I heard was that a DOD person overseeing research was a friend of Montroll and engineered the ONR fluid mechanics grant to LJI. The ONR scientists didn't like funding decisions imposed upon them, and the LJI grant was up for renewal. I went with an LJI team that came to brief ONR located then at 800 N. Quincy St in Arlington, Virginia, to seek further funding. In the back of the briefing room was a somewhat overweight man who slept through most of the briefing. Not a good sign. He was Bill Condell, the long-time head of the ONR Physics Division. The grant was not renewed, but later, through Bill, I was hired at ONR. Here is the story that shows how a number of turning points make up a career.

In 1983, Ed Ott and James Yorke, professors at the University of Maryland, held a nonlinear dynamics conference. Yorke was the person (together with T. Y. Li) who in 1975 gave chaos its name in the paper *Period Three Implies Chaos*. Ott became, perhaps, the most prolific high-quality contributor to nonlinear dynamics mathematics and physics. Getting back to the Ott–Yorke meeting, Bill Condell came and was introduced to me by Elliott Montroll, who praised my scientific work. Bill said, come to ONR and create a nonlinear dynamics program. I declined and said I want to work as a scientist, not as a program manager. But I did apply and win, through Bill, an ONR grant to organize a large international meeting at NBS on the topic of fractal random processes. Shortly afterwards, we got a new DARPA program manager, and it became clear that our DARPA grant would not be renewed. I called Bill and said I reconsidered and will apply for the ONR position. I got the job, as Bill was the deciding official. The government offer was for a GM-14 rank with a $45K salary, based on my $38K LJI salary.

By the way, when I graduated from Stony Brook in 1970, I interviewed for a position with NASA Goddard, in Greenbelt, Maryland, and was offered a GS-7-level position. It was better to get the PhD before going for a government career, as the government offer is tracked to your salary while promotion within the government can be at a lower salary.

I started in September 26, 1983 (and stayed until January 12, 2024). That November 1983, I organized the NBS fractals meeting with the ONR and other funds. If I stayed at Maryland I would have been stranded, out of a job, as the DARPA funds would run out, and sadly, Elliott Montroll died that December. I did have another opportunity. Robert Hart, a former Montroll student, was the head of basic research at the Johns Hopkins Applied Physics Lab (APL) in Laurel, Maryland, and he offered me a job. I chose ONR, but stayed in close contact with APL and gave several colloquia there. The ONR choice exceeded my expectations by affording me a great platform to continue my research and endless opportunities in the world of science. In addition, I quickly moved up in 1986 to a GM-15 position and in 1987, a Senior Executive Service (SES) position.

10

Before the Mast: A Decision to Stay at ONR

Elliott Montroll retired from the University of Rochester and moved to the University of Maryland where I joined him in August 1980. One day, he came back from a trip to his company, the La Jolla Institute, with an amusing story. He met with a psychiatrist, Arnold Mandell, the Chair of the Medical School's Psychiatry Department at UCSD. This man, Arnold Mandell, wanted to learn the math and physics of dynamical systems. He was studying the release of growth hormone, and its time series result was baffling. Typical measures such a mean, wandered, and never settled down. He believed he was seeing chaos and wanted to learn more about it. He also believed this new mathematics of complex systems would have an eventual relevance to neuroscience and psychiatry. At the conclusion of Mandell's one-hour meeting with Elliott, in his medical tradition, he paid Elliott a $100 fee as if Elliott was the psychiatrist and he was the patient who came to learn mathematics. Maybe it was from his pocket, or maybe he billed it to a grant as a consulting fee. Anyway, Elliott thought this was hilarious, a psychiatrist paying him.

Once at a European nonlinear dynamics conference, a psychiatrist spoke about recording his mood when he awoke each morning. He stressed that it looked like a chaotic behavior with no well-defined mean. I wake up each morning the same, so when I met with the speaker afterwards, he thought his mood pattern was normal and mine was weird. Maybe his is normal for psychiatrists? After working many years with Mandell, he said he would have liked to switch nervous systems with me.

Soon after I joined ONR in 1983, Mandell called me and asked to meet as he would be coming to the DC area, I guess to NIMH. We met for dinner at a Korean restaurant, and he proposed that my nonlinear dynamics program could become a unique nationally leading program. My initial thought was to stay at ONR for a short time to learn the grant business from the inside, but after that dinner with Arnold, I took up the challenge to make the ONR nonlinear dynamics program a national treasure. Arnold and I became close friends, and we started working together on protein design, and I ended up staying at ONR for 40 years.

Years later, Mandell won the MacArthur "genius" award to study dynamical systems and their relevance to neuroscience. Mandell wrote a book, *Coming of Middle Age: A Journey*, that included an account of his growing up in Chicago. On the McArthur board was the writer Saul Bellow, who from age 9 grew up in Chicago, and Mandell surmised that this played a role in his award selection, based more on literary than scientific talent. Sometime later, Mandell nominated me for the McArthur Award and got a note back from Murray Gellman thanking him for the suggestion, but nothing came of it. I invited Arnold as an after-dinner speaker at some of our dynamics meetings. Being a neuro-psychiatrist, he once spoke on the EEG of a university president vs a college student in terms of the signal's fractal dimension. This is the type of speaker that you want. Another time, Arnold gave the dinner talk at Café Mozart in DC in honor of Bill Saenz, who was retiring from NRL. We were at a long table on the side and not as expected, in a private room. Arnold gave a brilliant exposition on the possible connections between dynamical systems theory and brain wave states. When he reached the part about erotic dreams, the manager asked him to wrap it up as the patrons stopped eating and ordering and instead were rapt listeners. In 1986, we held NIH's first meeting on the role nonlinear dynamics could play in health. I invited James Gleick, a science writer, to our conference with proceedings published as *Perspectives in Biological Dynamics and Theoretical Medicine* (Annuals New York Academy of Sciences, Vol. 504, 1987), and he partially covered it in the bestseller book *CHAOS*.

When I joined ONR in September 1983, Bill Condell had already stepped down as the Physics Division Director. I was the new guy among seasoned program managers, one of whom was just promoted to the Division Director position. Our position was labeled as Scientific Officers, giving us a hint of a military flavor. Physics was one of the four divisions in a Science Directorate. ONR had three other directorates: Oceanography, Electronics, and Biological/Medical. All supported Basic Research, labeled as 6.1 funds in the DOD budget. In 1986, the Science Directorate Head, Ted Berlincourt (a physicist), moved over to the Office of the Secretary of Defense (OSD) to oversee 6.1 research. Our Physics Division Director was named Acting Directorate Head. He appointed me to be the Acting Physics Director. I suspect he wanted someone who was least qualified, and if he had to return to the post, it would be welcomed.

As an aside, Ted Berlincourt started an annual meeting for all science division heads from AFOSR, ARO, and ONR. We met at NRL, and Ted's instruction was "tell me three interesting things". After Ted retired, this meeting became the annual

TARA (Technical Area Research Assessment) review for all DOD science. I represented ONR and NRL physics, and an outside advisory board was set to review our detailed review presentations. In 2000, I even received a Special Act Award for my TARA work. This review became a major exercise. Physics was spread throughout NRL, and I had to try to get input from various research groups. The quickest helpful response was always from Jerome Karle, a chemistry Nobel Laureate. Some groups never replied. Eventually, the TARA became such a big, time-consuming, and expensive endeavor with little impact, that this growing bureaucracy was ended.

Back to ONR. The Executive Director of ONR was Marvin Moss, a physicist. He and several other ONR people had worked under John Deutch when John was at the Department of Energy, eventually as the Under Secretary. My first week at ONR, I learned that John would be heading an external review panel of ONR next week. So, I called him at MIT and told his secretary that Dr. Shlesinger would like to speak with him. He answered the phone, Yes Sir, how can I help you. He thought I was James Shlesinger, who had been Secretary of Defense and Energy. I said, no it's Mike Shlesinger. He said, Oh shit. When I said I joined ONR that week, he invited me to come up to Moss's office at noon the day of the review. I came up and John said, hello Mike. He then turned to Marvin and said, why didn't you say hello to Mike. Marvin said, who's Mike, and John then berated him with Mike is the best person at ONR and you don't know him? I interrupted with I just joined ONR on Monday and John enjoyed the gag. Marvin and I became friends after that.

Marvin wouldn't approve the Acting Science Directorate person for the permanent position, so I stayed as Acting Physics Division Director. When, in 1986, Marvin left for the Scripps Institute of Oceanography, Fred Saalfeld, who was the Research Director, was promoted to be the Executive Director. The Science Directorate position moved from Acting to Filled, and

I became its Physics Division Director, a Senior Executive Service (SES)-level position, and I decided not to move back to the University of Maryland.

The SES started under President Jimmy Carter, and it was fashioned as the equivalent of General and Flag Officers in the military. The Navy roughly had one SES for each Admiral. Prior to the SES, the civil service had grades GS1–GS15, and super grades GS16–GS18. People had a choice to stick with their super grade or switch to an SES grade. The SES position did not have tenure, and like a military officer, one could possibly be reassigned to different executive positions and locations, perhaps, every 3 years. Many thought the SES would disappear after Carter and they stuck with super grades. ONR and NRL were able to sweep up dozens of SES positions. While a Warfare Center with 2500 people might have two SES positions, I at ONR as an SES had around 10 people under me. We included all the people funded (PI's, students, and staff) as being counted as under our command, so on paper, it looked like I managed hundreds. Over time, the SES slots equilibrated with ONR losing many with ONR Division Director positions no longer being SES. But, in 1986, it was attractive, so I stayed at ONR and did not return to the University of Maryland. I spent a month at the Federal Executive institute in Charlottesville, Virginia, completing the Executive Excellence Program for incoming SES from all Executive offices. Regular GS positions could only carry over 240 hours of vacation time. One advantage of the SES position was that 720 hours of unused vacation time could be carried over at the end of the year. I had saved up that much, and during COVID, as I mostly did not use vacation time in my final year. I was able to cash in over 800 hours of vacation time when I retired on January 12, 2024. Also, an SES could be part of a larger bonus pool.

Today, ONR is a large complex organization with roughly $2.5B to manage from basic research to future naval engineering capabilities with prototypes and testing. Some managers adapt the carrot and stick approach and make work and competitions stressful while they had an eye on maximizing their year-end bonus. Most, however, are supportive and appreciative, the category that I put myself in. I've heard the phrase used about changing positions and managers in some cases, as moving from Sparta to Athens. ONR provided so many interesting opportunities and support to follow them that I stayed for 40 years, a biblical number alluding to a long time. All memories are present at a single time, what in general relativity is called space-like. I never felt that my time at ONR was long. It seemed like I was exploring space with time standing still. So, in relativity physics terms, my ONR experience was space-like and not time-like.

As a grad student, I thought my career path was to follow in the footsteps of my PhD advisor as a tenured professor, teaching and mentoring graduate students. Of course, I would need to win grants to support this ideal. Being an ONR Program Officer seemed orthogonal to that dream. How wrong I was!

When I joined ONR, I had published 38 papers, presented 48 lectures, and organized 1 conference. By the time that I left ONR, I had published 215 papers, presented 400 lectures, took many trips annually, organized or co-organized 40 conferences/workshops/symposia, published a mathematical memoir, co-founded the journal *FRACTALS*, co-edited 18 proceedings and 2 issues of the *Naval Research Reviews*, awarded 3 patents, was a PRL Divisional Associate Editor, was on 7 Editorial Boards, consultant at General Electric and Exxon, a two-year Chaired Professorship at the USNA, visiting Professor at the University of Melbourne, Blue National Command Authority at the Naval War College's Sea Conflict War Games, Physics panel member of the US–Israel Binational Science Foundation, among many other activities. In addition, I received a number of awards. Basically, ONR supported and afforded me any opportunity that

I sought, basically letting me design my career. Some of my professor friends thought I had the better deal and filled me with complaints about stresses being at a university.

All of my successes were due to having wise supportive leadership and teamwork. ONR has around 130 Program Officers with an array of different specialties and expertise, in approaches to science and engineering. ONR had so many people to learn from and to join in teams for multidisciplinary programs. I worked with Tom McKenna, a neuroscience expert, and Steve Ramberg and Tom Swean, first-class oceanographers, and Reza Malik-Madani and Bezhad Kamgarparsi, mathematical experts. Some were experts on engineering and some more experts on basic science. Fred Saalfeld, a chemist and the legendary Executive Director for 20 years, was knowledgeable about all aspects of ONR and a true leader. Bruce Robinson, a physicist, was a brilliant Director of Research. Gene Silva, an oceanographer, was a source of wisdom and encouragement. He funded indirectly the discovery of the Titanic. The list could go on for quite long.

11

History of ONR

"There was no government money for physics before the war"
(quote from I. I. Rabi, Physics Nobel Laureate, from 1979
lecture with war meaning WWII)

By the way, Rabi's initials I.I.R. written as I^2R represent resistive loss in an electrical circuit. Some said this also represented the loss in trying to follow a Rabi lecture.

The following tells the story about how government funding was initiated with the Congressional creation of the Office of Naval Research.

The following is a report I wrote, but never published, about the ONR 50th Anniversary. ONR is in 2025 in its 79th year.

The Office of Naval Research
Fifty Years of Excellence

When Planning, Forstering and Directing Scientific Research at the National Level was Needed: The Navy Stepped Up to the Challenge.

The Origins

The idea for the Office of Naval Research (ONR) was initially presented in September 1944, by a small group of young Naval Officers, as a beneficial suggestion for the Secretary of the Navy. The path from there to ONR's creation was neither easy nor straightforward. But, it was successful.[1,2] ONR was established on August 3, 1946, following President Truman's signing of Public Law 588 which was passed by the 79th Congress on August 1, 1946. Awarding and managing research contracts was a radical idea for any government agency, as much as receiving funds was for universities. But the concept was in line with Vannevar Bush's vision presented in his 1945 report to the President entitled *Science the Endless Frontier* which called for the continuation of the partnership between government and scientists developed during World War II.

The first Chief of Naval Research was Vice Admiral Harold G. Bowen. In 1939, as Director of the Naval Research Laboratory, he had already started a program to investigate the feasibility of nuclear-powered ships employing the energy of nuclear fission. Such groundbreaking scientific thought was in the Navy tradition. Naval officers helped found the National Academy of Sciences, and in 1907, Albert Michelson won the first US Nobel Prize in Physics for his experiments, begun at the US Naval Academy, discovering that the speed of light was a universal constant. This tradition continues with Jerome Karle's 1985 Nobel

Prize in Chemistry for his work at the Naval Research Laboratory on molecular structure. NRL was founded in 1923 at the suggestion of Thomas Edison and today plays an integral role for ONR as its world-class corporate laboratory.

Like his successors, Admiral Bowen was granted broad powers to encourage, promote, plan, initiate, and coordinate Naval Research. Speaking to the Navy League in Schenectady, on October 10, 1946, he said, *Scientific research and development has reached a point in our lives, where to ignore it or even be casual about it would be folly.* He went on to encourage *youth to pay more attention to science, not only in college but earlier in the secondary schools.*[3] Fifty years later, ONR can proudly point to generations of scientific research and students that it has supported.

Navy Capt. Robert Dexter Conrad, first Director of the Planning Division of ONR, designed the ONR organizational structure and the university contracts program. He traveled around the country explaining ONR to the university community. An oceanographic research vessel was named in his memory and today he is honored by the Conrad Award presented for scientific research and leadership in Naval Research. Speaking at the University of Illinois on October 27, 1946 he said, *The Navy has embarked upon a venture which is not only new to the Navy, but new to the Government and which is of deep significance to the national welfare as well as to the national security.... it is clear that it is a contradiction to speak of directing and controlling research. An unexplored country cannot be mapped. It is proper and necessary to plan development work, but research must follow only its inner promptings.... Atomic bombs, guided missiles, bacteriological poisons, and all their hellish brethren and potential offspring have created vast fears and doubts in the minds of men. The old securities of space and time are vanishing. Our powers of self-destruction appear like a baited trap ... Where is the new hope? Where is the new security? The answer is in knowledge. The renaissance of research to which the Navy is proud to contribute, can create the new knowledge and stimulate education which are the foundations of a better world.*[4]

The first Chief Scientist of ONR was Alan Waterman, who later became the first Director of the NSF. In 1956, Waterman received the first Conrad Award. Upon accepting the award, he commented, *There has been an uneasy feeling on the part of many that the problem of the rising costs should be met in part by curtailing scientific research ... This would be a serious mistake. It is like saying that an industry should take prudent care of its future interests by failing to replenish its stockpile and sources of man power and curtailing its libraries and reference facilities. We must have scientific data, we must have ideas.*

We salute these early men of vision. Their words still ring out to us as longstanding universal truths. The early days of ONR are a bygone era of a simpler time of train rides to meetings, ships to Europe, and handshakes to close contracts. We have tried to keep that simpler spirit alive, sometimes with success.

The First 10 Years to the 50th Anniversary

ONR made its reputation during its first 10 years and set high standards for all who have followed, both in terms of scientific achievement and scientific management. ONR's well known London office, which helped rebuild European science, also began in 1946. (The Tokyo office didn't open until 1975). The original Science Branch of ONR had Physics/Nuclear Physics/Mechanics and Materials/Electronics and Communications/Math/Chemistry and Fluid Mechanics Divisions. There was also a Medical Sciences Branch and a Program Branch organized around surface ships, subs, air, power, armaments, amphibious, and geophysics divisions. Some of the research efforts of those early years,[5,6] listed in the following, hint at the excitement of the times: the Whirlwind computer in 1946 was the first magnetic core digital computer, and it led directly to founding of DEC; it was used in air defense systems; masers and quantum electronics research began which ultimately led to the laser; molecular beams physics provided techniques which were later used to build atomic clocks; superconductivity for cryogenic electronics later used in hydrogen bomb designs; nuclear physics for ship propulsion; boundary layer control for improved stability of high performance aircraft; towed arrays for submarine acoustic detection; balloon study of cosmic rays to investigate conditions for high altitude flight; shock waves area important for re-entry vehicles; and Atlantis I — an oceanographic research vessel. Other areas of focus included infrared detectors, superfluidity, radio astronomy, artic research, fluoride pastes to prevent tooth decay on long tours of duty, tilt wing aircraft, helicopter stability, pilotless aircraft to photograph hurricanes and weather forecasting, fluid mechanics study for designing hydrofoils, noise reduction for ships and machinery, underwater acoustics, blood plasma expanders, and tissue preservation.

Similar extensive lists of ONR activities can be constructed for the following decades.[7] In more recent times, thrusts in just the fields of physics and electronics include laser cooling of atoms and ions, Bose–Einstein condensates, acoustic holography, chaos, fractals, wavelets, neural nets, high power microwaves, atom interferometry, squeezed light, lasing without inversion, non-neutral plasmas in traps, thermo-acoustic refrigeration, long base-line interferometry, quantum chaos, free electron lasers, high temperature superconductors, clusters, buckeyballs, laser control of reactions, quantum wells, wide bandgap semiconductors, single electron transistors, diamond films, scanning tunneling atomic force microscopes, and quantum gates. The manner in which these investigations affect technology will afford much to expound upon at a future ONR anniversary. Now, in the present time of 2025, machine learning and artificial intelligence, hypersonics, metamaterials, plasmonics, chaos morphable logic gates, and quantum computing are some of the trending physics topics.

The Celebration

A grand meeting honoring ONR's 50 years of commitment to excellence in research was held at the National Academy of Sciences on May 22, 1996. The meeting was opened by Rear Admiral Marc Pelaez, the Chief of Naval Research. By this time, ONR had grown beyond its initial 6.1 research portfolio to manage the Navy 6.2 (Applied Research funds) and Navy 6.3 (Advanced Technology Development funds). The other speakers were John Gibbons (Office of Science and Technology Policy), Richard Danzig (Under Secretary of the Navy), John Douglass (Assistant Secretary of the Navy), Rear Admiral Richard Riddell (Director of the Navy Test & Evaluation & Requirements), Lt. General Paul Van Riper (Director Marine Corps Combat Development Command), Charles Vest (MIT Vice President), Robert Frosch (Kennedy School of Government), Robert Galvin (CEO of Motorola), and Frank Press (National Academy of Sciences-Past President). They reminded the audience of the wartime advances in code breaking, radar, and nuclear fission, which transformed thinking on the role of science in the defense of the nation. They also reminded that research budgets needed to be rigorously defended every year. The research return on investment was estimated to be in the range of 30–50% each year. Even back then the WEB was held up for its unexpected impact on research information and exchanges. One of the Young Investigator grantees, Frances Arnold, spoke on speeding up evolutionary changes in proteins. Her work won the 2018 Nobel Prize in Chemistry.

The meeting closed with Fred Saalfeld, the ONR Executive Director, speaking on how the partnership between government, academia, and industry accomplishments has fulfilled, and gone beyond, the promise of 1946 ONR.

The ONR Program Officer

The key element in ONR is the Program Officer. I was hired as an ONR Science Officer. Sometime later, this type of position was changed to Program Officer. One had to get used to changes in the DOD. Over time, I was in three different ONR Departments and four different Divisions. The main job of the 130 Program Officers is to learn, think, create, decide, and defend research programs. The Program Officer must create national programs in science and engineering which are a combination of first-rate cutting-edge science directed toward technologies that can enhance the defense of the nation. The programs need to mix long-range visions with shorter term goals, make or break paradigms, blend low to high-risk efforts, and balance moderate against high payoffs. The Program Officer has to know how research actually occurs and ensure the continuity of work. Each ONR proposed program must be won internally through stiff competition. Successful programs are reviewed annually, but once won, it is the ONR scientist and engineer who decides which proposals to fund, unlike, say, the NSF, where outside reviewers rate proposals. So, hearing an exciting talk at a conference, the ONR Program Officer can approach the speaker and start the process to get a proposal and award a grant. No need to wait for outside reviews and internal meetings for funding decisions. A Program Officer funds innovative, leading-edge research that, many times, is in a direction which does not mirror the present (but hopefully the future) consensus of the scientific community. It is the Program Officer who visits your lab, meets with your students, organizes exciting workshops in your field, and is the first one you feel like calling when you make a breakthrough. The Program Officer is the heart of ONR.

This article is dedicated to the 1944–1946 efforts of LT Commanders Bruce Olds, Ralph Krause, and H. Gordon Duke, and LTs James Wakelin and John Burwell, who proposed and invented what became the Office of Naval Research.

1. H. M. Sapolsky. *Science and the Navy: History of the Office of Naval Research*, Princeton University Press, Princeton, New Jersey (1990).
2. S. Bruce Old, Physics Today. *The Evolution of the Office of Naval Research*, (1961).
3. *ONR's ORIGINATOR*, 2(73), (1946) ONR NLD.
4. *ONR's ORIGINATOR*, 2(84), (1946).
5. *Naval Research Reviews*, 48(1), (1996).
6. L. M. McKenzie. After six years — A study of the impact of the physics branch program. *Science*, 118, 227 (1953).
7. *Naval Research Reviews*, 47(4), (1995).

The Admirals: Chiefs of Naval Research

By the law establishing ONR in 1946, it was designed to be led by an Admiral, the Chief of Naval Research (CNR). Promotion in the military is by merit. When a person becomes a Flag Officer (that is, an Admiral), they have accomplished much in their career and proven themselves. There were fourteen previous CNRs, and then when I joined, I served under fourteen CNRs: Admirals Kollmorgan, Mooney, Wilson, Miller, Pelaez, Gaffney, Cohen, Landay, Carr, Klunder, Winter, Hahn, Selby, and Rothenhaus. With Kollmorgan, I only overlapped for a short time at the end of his tour, and with Selby, due to COVID and work from home, we never met. Adm. Miller had an engineering PhD degree from Stanford University. After ONR, he was the Provost at the USNA and approved my two-year Kinnear Professorship at the USNA. Adm. Pelaez had a love of mathematics, and some days after work, he would come to my office to program nonlinear equations with me. Adm. Gaffney was a well-known oceanographer. To a great extent, the Admirals represented ONR to the Navy and defended our direction and budget, while the ONR Executive Director was more involved with working inside the building. When I started at ONR, I was in my 30s and the Admiral was probably in his 50s. When I retired, I was in my 70s and an Admiral was still in his 50s. ONR was, rock solid, a military institution, and when the Admiral decided something or needed something, then that was an order and took top priority. Under the Admiral, there is the Assistant Chief of Naval Research (ACNR, a Navy Captain) and a Vice CNR (a Marine Corps General dual hatted from the Marine Corps Combat Development Command in Quantico), and each research department has a Military Deputy Department Head, usually a Navy Captain or a Marine Corps Colonel. When tasks come down the chain of command, the military officers always took each one seriously and responded accordingly. The scientists less so. Some Department Heads decided whether or not the task was critical and only then passed the work further down to the scientists. Some of the requests were vague. As a Division Director, I would input my response to my fellow Program Officers so they would have an example to follow. It was not uncommon, once all the departments responded that the task would need to be reworked with a new directive, we like the way this person responded, so everyone re-do your response in that style.

Once, Admiral Mooney came back from the Pentagon embarrassed. I think he was meeting with other Admirals connected with the Chief of Naval Operations and the Secretary of the Navy. Someone asked him "what's new", and he wasn't prepared to answer. On returning to ONR, he instituted a policy that every Division each week would provide a vu-graph of something new, so he would be prepared to answer that question. Well, there were, at that time, I think fourteen divisions, so after some time, the number of vu-graphs grew. Once I had to represent ONR at a meeting and present some accomplishments, so I went to the CNR office to get the CNR vu-graphs. There were filing cabinets full of them, and apparently, they were untouched. After Adm. Mooney, the practice was stopped. As related elsewhere, the TARA review processes became so overwhelming with generating presentations and bringing in review panels that it was discontinued. So, the scientists took each task with a grain of salt, but the military officers took each one very seriously.

I had a discussion with a Naval Officer. I said the reason the Manhattan project worked is that they got the best scientists and gave them all the resources they needed. He had a different perspective. He said it worked because they put a fence around Los Alamos with armed guards and told the scientists that they could leave only when the project was successful. When I headed ONR's counter-IED program, I wondered if success would come more quickly if scientists were set up for 24 hours a day in an enclosed guarded town and only allowed to leave when successful.

Getting back to Los Alamos, the decision to build an atomic fission bomb was daring. One needed U-235 which was rare and had to be separated from the abundant U-238. The Germans greatly overestimated the quantity of needed U-235. Rudolph Peierls and Otto Frisch, in England, realized neutrons could bounce back from the explosive boundary to multiply the fission effect, and from this, the Los Alamos project proceeded. In any event, it would be daunting to come up with kilograms of U-235 for a single bomb, and it was a daring judgment to go ahead with the expensive project. As an aside, I did meet Peierls in Sitges, Spain, in 1974. A brilliant lecturer who spoke, in part, about his 1929 concept of the Umklapp phonon. He and his wife, Genia, were friendly and generous with their time with the grad students. Research at Berkeley found that slow neutron captures could lead through a double beta decay for U-238 to be transformed into plutonium which was fissionable. The Los Alamos gamble paid off. ONR's over 100 Program Officers also take risks with high payoffs.

The Erdos Way

There can be alternative ways for the advancement of science and mathematics. I learned of one while attending a seminar, at the University of Rochester, by Paul Erdos. He is a renowned Hungarian mathematician with an unusual career. Later in life, he had no affiliation but rotated among mathematician hosts. His seminar was a collection of unsolved problems, including in graph theory, probability, and number theory. If I remember correctly, he started with drawing at random m lines and randomly placing n dots, then asking questions about how many dots were enclosed by different sets of lines. Then, he offered $100 to anyone who solved the problem. People took careful notes. Then, he moved on to progressively harder problems and larger offers of remuneration. I think he ended with a $10,000 reward problem. Through many seminars, he enticed dozens of people to work on his unsolved problems. He typically co-authored with those who succeeded in finding solutions, and he ended up with over 1500 publications and ran an international network of mathematicians at the cost of only maybe $10,000 a year.

What if ONR only held competitions on unsolved problems and only paid for success, instead of funding thousands of grants? DARPA does have challenge programs, such as one for driverless cars completing a course, with winners being awarded funds. Maybe the ONR way mixed with the Erdos way would be a winner.

CHAPTER 12

The View from the Other Side of the Desk

Typewriter and No Cameras

When accepted to college, you look at the school catalog to peruse the course offering. Your view is that college is about teaching (and maybe sports team rankings). In graduate school, you think the university is about research and that your thesis advisor knows all, and you will be given a good research topic. No matter how many PhD students, the advisor has good topics for all. Well, my advisor thought one needed to find their own thesis topic. When you are a professor, you teach, do research, train students, hope to get a summer salary, and then have to find funding to support all of that. Ah, the graduate student days seemed so much better in retrospect. As a professor you sit on your side of the desk trying to convince a funding agency about the merits of your research. My side of the desk is less stressful. Much easier to choose who to fund than to win funding.

What do I look for in a proposal that covers a relevant topic? I need to be able to master the material in a proposal and be able to defend it in internal reviews without notes. Therefore, it would be helpful to have a proposal be well written and include several sections. Starting with an executive summary that upper management could read and understand is essential. Then, a short technical abstract spelling out the problem, approach, and payoff. The proposal could start with a history of the topic, where the topic stands today, and what obstacles to progress are. Then, what you want to accomplish and what your approach to overcome any obstacles is, in detail. What your progress could lead to in terms of new topics or approaches.

Seeking grants, during my time at the University of Maryland, I had been to ONR, NSF, and DARPA to meet program managers. When I joined ONR, I thought my whole day would be scheduled to meet grant applicants. I was completely wrong. Even back then in the 1980s, hardly anyone showed up in person. These days, grant proposals are submitted online on grants.gov. Submitting a grant proposal to me was a level playing field. I had not worked on nonlinear dynamics and had no connection to the field, so I had no bias or personal connections to sway my funding decisions. Some grant applicants to ONR were from people whom I met at conferences and site visits to universities. So, typically, no walk-ins; therefore, no one was sitting on the other side of the desk. Anyway, if you came to my first ONR office, it was an interior room with no windows. One of the other Scientific Officers, as a joke, taped a picture of a window to a wall in my office.

In reality, even if one was doing great innovative work and wrote a great proposal, that does not assure funding. I had a budget that only allowed so many grant awards. Also, my program had a theme of nonlinear physics that could cover a range of topics involving lasers, semiconductors, or any system that can take advantage of nonlinear behaviors. If I could not fund a great proposal, I would keep it on file and tell the proposer that if additional funds become available, or there is an increase in next year's budget, I will try then to bring it into my program. I tried to be encouraging and helpful, that is being a public servant.

Much later, one day, we were all told our positions were changed from Scientific Officer to Program Officer. I preferred the former title. Initially, when I joined ONR, there was an information desk in the lobby, and anyone could walk into the building during office hours. Eventually, that changed into a security desk with signing in and a need to be in the building. Occasionally, I did get visitors who needed to sign in and be escorted. I would go down to the lobby to escort visitors back to my office. My security badge swipe could open the entry gates. My badge also worked to enter the Pentagon. If someone was not an American citizen, there were a few extra steps to get them into the ONR building. In those cases, it was sometimes easier to meet outside in the Ballston Mall across the street, so no sitting across from my desk. NSF and NIH also changed to going through security to gain entrance. NIH erected a metal fence around the entire campus.

In 1983, on the office desk, we had a typewriter and a telephone with buttons that could pick up calls on any of the lines for the other physicists. When I became the Physics Division Director, I used to stay at my desk from 12:00 to 13:00, so if anyone called the government, someone would be there to pick up the phone, me. I thought if you went to the post office, someone would be there, and if you called ONR, someone should answer. Eventually, that was replaced by voicemail and then email and smartphones. In later years, I never got landline phone calls. We even emailed each other to see who was going to lunch. Our first desktop computer was, I think, an IBM 286, and attempting to do grants on it was challenging. Slowly, it was being replaced by the IBM 386, but the older 286 computers were disappearing overnight. As a preventive measure, the cleaning crew that worked after office hours were told only to work during business hours, as if they were the culprit. So now, during the day, one was interrupted by vacuuming and trash removal. It turned out it was a security guard on the night shift doing the pilfering, one computer at a time, and not our friendly cleaning crew. In any event, if not for the guard, someone would need to remove the old IBM 286 machines.

One early rule, for security reasons, was that no cameras were allowed into ONR. Anyway, once everyone had a smartphone, no one seemed to care that everyone was bringing a camera into the building. If anything was classified, it was in a safe and not sitting around in the open. Occasionally, I would be sent some classified secret reports. I needed to sign them out of a safe and sign them back in. None of them really ever looked worthy of being classified. I was once in charge of administering an ONR grant to the National Academy of Sciences for a study of how to counter improvised explosive devices (IED). These were roadside bombs, booby traps, etc. that were hard to discover and caused casualties. The NAS final report was somewhat based on interviews with scientists whom we were funding. The NAS classified their report, although the work ONR funded was unclassified. The upshot was that I couldn't send or share the report to anyone unless they had a secret clearance. I could hear things at the classified meeting that I already knew, but now I couldn't discuss those things openly. I guess my view is that too much is classified. Maybe people classify reports to make their work look serious. Years ago, in the 1980s, when I consulted at Exxon, people mailed everything via Federal Express. I guess to make it look that their correspondence was too important for the mail.

Theory or Experiment

Starting in the summer of 1968, the end of my sophomore year at SUNY Stony Brook, I began 2 years with the nuclear physics Van de Graaff Lab. I learned what it takes to build, maintain, upgrade a lab, wear a radiation badge, and plan and carry out research projects. Groups scheduled beam time and worked around the clock when it was their turn. My job, besides optimizing the beam and running the accelerator to collect data, was mostly programming the new PDP-9 computer for data analysis. When the PDP-9 was unpacked, I was the first one who read the programming language instruction book and the first to start programming on it. I was also converting FORTRAN programs written for the IBM 7044 to be converted to run on the new IBM 360. This meant converting from 36-bit coding to 32-bit coding. One complication was the 7044, for alphanumeric characters, which used Binary Coded Digits (BCD), needed changing to Binary Coded Decimal Interchange Code (EBCDIC) used by the 360. This lab time for a future theoretical physicist gave me a great appreciation for experimental physics. Maybe all physicists, at some point, should work in a lab.

When briefing my programs at annual ONR physics reviews, it was easier to describe experimental work. The review would be to upper management and an outside panel. For an experimental project, I would show a picture of the equipment and data plots. I wouldn't add much to the slide, except a headline; that way, the audience needed to rely on me to explain the slide. If too much information was on a slide, the reviewing audience would get bored and flip ahead. These were the days of vu-graphs and paper handouts. The review panels comprised scientists from a variety of physics disciplines. Presenting a theoretical work can easily go beyond the knowledge and patience of the audience, as ONR covers so many areas of science and management cannot be conversant in all. So, to explain a theoretical work was much more difficult when not in a panel member's field of interest. Plus, I also needed to understand the theoretical work clearly enough to explain it and answer any questions. My program tended to favor experimental works.

At the University of Rochester's Strong Memorial Hospital, I went to a 50th year anniversary conference in 1976. One speaker was Alan Hodgkin, a Nobel Prize winner, for the Hodgkin–Huxley experiments underlying the model of ionic Na and K currents through neural membranes. During the question period, one person said there were two theories about some point in the talk and which one he favors. His response was sharp, do the experiment to find the answer, that's better than theorizing! For biology, Hodgkin certainly sided with experiment over theory. It reminded me of my high school biology teacher remarking one day, why would anyone go into physics, it's just equations. Today, I believe some of the best physics topics are about biology on all levels, from molecules to cells to organs to life, experiments, equations, and all.

I wish to point out a few of the people that I was honored to fund through ONR. They were all exceptionally talented scientists and true leaders in their field and mentors to their students. They worked on a variety of topics, and ONR support was able to fund some of their efforts. All the Principal Investigators that I funded made me look good. First, the experimentalists.

There were several experimental groups at universities that I funded through ONR that were exceptional in mentoring graduate students and postdocs while producing advances in physics, and anyone would be fortunate to join such a group. ONR only partially funded their work as they had several research topics and associated experiments. For example, at Purdue, Vlad Shalaev and Alexandra Boltasseva, lead a group on the physics of plasmonic nanolasers; Chee Wei Wong, at UCLA has a group working on micro-resonators producing dissipative Kerr solitons, and tunable frequency combs; there's Hui Cao's group at Yale on random lasers; Mercedeh Khajavikhan's group at University of Southern California on quantum exceptional point physics where real and imaginary eigenvalues meet for enhanced detectors; Dmitry Budker's group then at Berkeley and Michael Romalis' group at Princeton on novel magnetometers, Harry Swinney's U. Texas/Austin rotating fluid experiments on chaos, coherent structures, and fractal stochastic behaviors, Robert Westervelt's group at Harvard when I funded him for early work on semiconductor electron-hole chaos; and Raj Roy's group at the University of Maryland on optical synchronization, and Ken Showalter at West Virginia U on coupled chemical oscillators. These leaders are true mentors, and by the way, if you remember, Mentor was the wise advisor to Odysseus in the *Odyssey.*

In the same vein, there are some great theory groups that I funded on a variety of nonlinear physics topics including Ed Ott at the University of Maryland (see his wonderful book, *Chaos in Dynamical Systems*), Ying-Cheng Lai at Arizona State University (see his important book with Tamas Tel *Transient Chaos*), and Henry Abarbanel at UCSD (see his novel book, *The Statistical Physics of Data Assimilation and Machine Learning*). Then, there is Marlan Scully, at Texas A&M U (see his fundamental book with Suhail Zubairy, *Quantum Optics*), who has an exceptional group that excels in combining theory and experiment in quantum optics. Other great ones who have since passed away were George Zaslavsky at the Courant Institute, Leo Kadanoff at the University of Chicago, Steve Orszag then at Princeton, Harry Suhl at UCSD, Frank Moss at U. Missouri St. Louis and Mark Stockman at Georgia State U.

I steered away from large, diverse theory groups where the Professor co-authored more papers than was possible if actually contributing. During my graduate student days in Elliott Montroll's group, grad students would add his name to papers, and he would remove it. He only authored papers to which he actually substantially contributed. As in Montroll's case, when I checked online for great physicists, their publication records were on the order of 100 papers. At ONR, I could receive a proposal where the PI could have over 1,000 publications. Too overwhelming for me.

The Experimental Chaos Conference

My ONR program helped develop nonlinear mathematics with a rich array of concepts, such as strange attractors, strange kinetics, chaos, and fractal dimensions. The program also discovered physical phenomena exhibiting these behaviors. For example, the first chaos control experiment was done at NSWC, on a driven magnetoelastic ribbon (Mark Spano and Bill Ditto). The first chaos synchronization experiment was done at NRL (Lou Pecora and Tom Carroll). ONR, along with NRL (Lou Pecora and Sandeep Vohra) and NSWC (Mark Spano and Bill Ditto), we created, in 1991, the popular biannual Experimental Chaos Conference (ECC). The first was held in Crystal City, Virginia, two Metro stops from the Pentagon. It was a huge success, and the second meeting was also held in Crystal City. Both proceedings were published by World Scientific Publishers. It became international with the 3rd conference in Edinburgh. The 14th one was held in Banff, Canada, in 2016, and the 15th in Madrid in 2018, and in 2021, the 16th in Xi'an, China. It is now known as Experimental Chaos and Complexity Conference.

In 1900, in Paris at the International Congress of Mathematicians, David Hilbert famously created a list of problems that could occupy mathematicians for the new century. In trying to emulate Hilbert, I asked for ECC speakers to be experimentalists and to report which of their experimental results were in agreement with theory, what theory got wrong, and what results were not yet addressed by theory. I was hoping each ECC would provide a short list of open topics. The speakers were experimentalists, and the attractions were for theorists to attend to find open problems. As the list would grow with each conference, the list of solved problems would also grow. This approach was a complete failure as talks presented successes and not open problems. A list of experimental results needing an explanation was not generated.

Destination USA

When I started graduate school in Physics at the University of Rochester, about one half of my fellow students were foreign from a variety of countries. They all seemed to already have master's degrees and were more advanced than myself. My initial reaction was that I would need to compete with them for postdoc jobs, and in the mid-1970s, jobs were scarce. In the 1960s, research funds were plush due to the space race and the Cold War. At that time, university physics departments were expanding in faculty positions and in accepting graduate students, both US citizens and foreign-born. All that was in decline in the 1970s. Soon, my fellow students became friends and a support group. Many stayed in the US and contributed greatly to US science. Strong funding did come back. In the 1980s, there was increased defense funds to outcompete Russia, then it was for not losing our industrial base to Japan, and later to outcompete against China. Years later, my ONR program supported excellent researchers in the US who came from countries including China, Russia, India, Iran, Israel, Bulgaria, and even Cambodia and Cameroon. Federal funding of science has made the US attract the best from around the world, bettering US science and creating science opportunities for all. The proverbial tide lifting all ships. A payoff of government grants, indirectly, was bringing top-notch budding scientists to the US in the first place as graduate students and then staying to better US science. It's somewhat unfair to home countries to attract their best educated and brightest from around the world. These days, the US may not be as welcoming, to our disadvantage.

Grant Decisions

At ONR, the Program Officers make the decisions of which proposals to fund within their budget lines. Unlike at NSF and NIH, there are no outside reviewers for ONR. For example, at NIH, I was an outside reviewer for NIMH. I can hear a talk at an APS annual March meeting and ask the speaker to send me a proposal and fund it immediately. Or the opposite from attending a conference. A visitor to my office said he could determine if any equation would support chaotic behavior. That sounded impressive. When he gave a talk at an APS meeting, I was in the audience. After the talk, a remark from the audience said that his method was trite and would be useless for most equations of interest. The speaker agreed, and I was no longer interested and did not request a proposal. This shows the importance of not sitting behind a desk and instead being out with the scientific community at conferences or site visits to people, their labs, and their graduate students, and also to keep up with the journals. When I hear an exciting talk at a meeting or get a great proposal, ONR can start funding new topics very quickly, as no need to wait for outside reviews. I can give scientific approval, and then the proposal moves up the chain of command for higher up approvals. When ONR only had 6.1 basic research funds, each new grant needed to be defended in reviews, by the appropriate Division Director, at weekly meetings called the Program Council. Then, the financial office approves, if upon checking that the funds requested are available, and finally, the Business Office signs off if all the costs are allowable and all required forms and costs are in order to officially complete the award. For example, there is a required signed declaration that the university is a drug-free workplace (even though the student body might not be).

Someone told me that they won't give away too much in an NSF proposal for fear that a reviewer would steal their ideas. As ONR does not use outside reviewers, I need to defend, during reviews, all proposals that I gave scientific approval. So, I won't select a proposal if I am not comfortable in understanding and explaining the ideas and technical approach. So, in reading proposals, I needed help from the author. My ONR program was Nonlinear Physics, focused on how to predict, exploit, or avoid nonlinear behaviors to eventually improve technologies or create new technologies. Nonlinear physics could be found in most areas of physics, for example, semiconductors, lasers, fluids, and plasmas. I am not and could not be an expert in all aspects of physics. All that said, there are today at ONR many different types of programs and proposals. A typical grant is for 3 years with possible renewals. I wouldn't expect someone to perfectly predict what they would achieve or be working on 3 years in advance. I usually told a new PI, don't worry if you don't follow the proposal in detail, in fact I don't expect that. Most felt that was relief and not a common instruction from other agencies and program managers. As a working scientist, my sympathies are with the scientist.

When I joined ONR in 1983, I had a core program and could compete for additional funds under an Accelerated Research Initiative. ONR also had a Small Business Innovative Research program. DOD has seven Budget Activity Codes. ONR had what is labeled as 6.1 Basic Research funds. Around 1995, ONR expanded to also manage the Navy 6.2 Applied Research and 6.3 Advanced Technology Development funds (also 6.6 funds for Management Support). In addition, a Program Officer's core programs, there are these days many other programs, each with its own competition and proposal deadlines and requirements:

- Naval Research Enterprise Intern Program (NREIP)
- Multidisciplinary Research Program of the URI (MURI)

- Defense University Research Instrumentation Program (DURIP) of the URI
- DoD Experimental Program to Stimulate Competitive Research (DEPSCOR)
- Young Investigator Program
- DoD National Defense Science and Engineering Graduate (NDSEG) Fellowship Program of the URI
- Summer Faculty Research Program
- Faculty Sabbatical Leave Program
- Naval High School Science Awards Program
- HBCU (Historically Black Colleges/Universities) Future Engineering Faculty Fellowship Program
- HBCU/Minority Institutions Program
- Science and Engineering Apprentice Program (SEAP) (Run by ONR, funded by the American Society for Engineering Education)
- Science, Mathematics, And Research for Transformation (SMART) Defense Scholarship Program
- SBIR/STTR Programs
- Vannevar Bush Faculty Fellowship managed by OSD

The ONR website will have calls for specific research topics.

Also, through the ONR Global office, research programs and conferences can be supported around the world. In any given year, a Program Officer, in addition to their core program, can be involved in reviewing proposals for any of these programs and managing the winners of the above list of programs. Very easy to be busy.

My favorite program to compete and manage is the Multidisciplinary University Research Initiative (MURI) program which brings together a diverse group of mathematicians, scientists, and engineers to attack a multidisciplinary problem. The MURI winning group could be from several universities and might, as an example, comprise a physicist, a mathematician, a chemist, and an electronics engineer. The MURI award gives a person a 5-year funded opportunity to work with other disciplines, even with other universities, and with DOD laboratories. By all accounts, MURIs are very popular. Henry Abarbanel introduced a winter school with his MURI programs. Being in La Jolla, a winter school was popular. The first day would be tutorials for students. The next day for outside speakers in relevant fields but not part of the MURI. The last day would be the MURI scientists presenting their works. Very educational, instructional, and appealing to a wide audience of researchers and students.

In the 1980s, I was an early supporter of the newly formed annual conference of the National Society of Black Physicists. These meetings were held at Historically Black Colleges and Universities (HBCUs). They were well attended with good lectures and an emphasis on students with support to continue on with their education and careers. My visits to HBCUs always found the students to be eager to become physicists and the faculty appreciative of government support. I enjoyed presenting an overview of ONR to a science fair at Morgan State University in nearby Baltimore where the students were enthusiastic.

I also enjoyed being a judge at the International Science and Engineering Fair for high school students where ONR gave out scholarships to top display booths. Several government agencies, for example, the NSF, were also represented. Over 1,000 students participate and compete for around nine million dollars in awards. The students are award winners for local science fairs.

The Good Old Days

When I joined ONR, I was told that I just missed the good old days. ONR had recently began to do its budget request through the Program Objective Memorandum (POM) process along with the rest of the Navy. Previously, it was a simpler separate affair, separate from Big Navy. Decades later, I told newcomers that they missed the good old days. The reason is again budgets.

The National Defense Authorization Act (NDAA), that is, the DOD yearly budget, needed to be passed and signed into law by the President before the start of the fiscal year on October 1st. For many years, that was the case. The NDAA was passed by Congress around July 1st, and I could start writing the increments for ongoing grants, writing renewals for ending grants, and the write-up for new start grants. Within my budget, I choose what topics and who to fund. I need to write up a short work statement, the approach to solving a problem or advancing a field, and the payoff. I would get all of this prepared in our computer system, so on October 1st, I only needed to open a file and hit submit. That was the good old days.

For the past many years, a contentious Congress, split along party lines, not only did not pass a budget on July 1st, but they did not even pass one by October 1st, the first day of the government fiscal year. What they did pass was called a continuing resolution (CR) which provided partial funds, maybe for 1 month. When that elapsed, a 2nd CR could be passed. I might do partial funding of some grants with the initial amount of CR funds, but then I need to do a second or third action when more funds become available later in the year. Double the number of funding actions might now be passed to the ONR contracts office, so doubling their workload. Typically, a budget might not get passed until late December or even later. Once the DOD gets its budget, then the Navy finds out its numbers in a variety of categories, for ONR such as 6.1–6.3, SBIRs, MURIs, etc. Next, ONR needs to carefully go over the NDAA to find out what bills it needs to pay to support Congressionally mandated activities and to determine how much it can provide to each ONR department and activities, such as MURI, HBCU, SBIR, and DEPSCOR. The department heads then decide the funding level for each division, and the division director sets the funding level for each program officer for each of their activities. Once all these decisions are made, the funding numbers are entered into the funding lines for each Program Officer. Once all this is done, time has elapsed, and it is now probably March when I get access to my full funds. Now, when I submit my grant and contract increments, renewals, and new starts to our contracts office, so have all the other hundred or so Program Officers, swamping our contracts office. My actions might not get finalized and funds made available to Principal Investigators until the summer. Remember, the PIs wanted to bill for the funds on October 1st. When the PIs submit their first billing for the fiscal year, it is now close to the end of the fiscal year, which is September 30th. Due to the Congressional delay, only a small amount of that fiscal year's funds has been expended. On October 1st, the Navy Comptroller and also Congress want to know why ONR has left over funds from the previous year. Obviously, everyone knows, but the game goes on. It is not clear to me if Congress knows all the complications it creates by not doing its job in a timely fashion.

13

My ONR Programs

No Two Days the Same

Feigenbaum's discovery of the universality of the period doubling route to chaos was a surprise. The shape of the map was important and not the specific equations. His paper set off an explosion of work in nonlinear dynamics. As was said about the Schrödinger equation was now said about Feigenbaum's work, that now second-rate people can do first-rate work. My interest in nonlinear dynamics was directly associated with Feigenbaum's period-doubling universality discovery.

As recounted before, I started ONR's Nonlinear Physics program in September 1983. Much progress had been accomplished before that. Since 1980, Los Alamos has an active Center for Nonlinear Studies (CNLS). They held an important international conference *Order in Chaos* at Los Alamos in May 1982. In addition to chaos, CNLS also studied the nonlinear equations for solitons, a localized wave that retains its shape after collisions with other solitons and is mathematically braced by an infinite number of conservation laws. My ONR program only supported soliton work much later with Al Osborne's inverse scattering transform technique to decompose a wave into soliton and radiation parts. Al has a great comprehensive book on all of this: *Nonlinear Ocean Waves and the Inverse Scattering Transform.*

The ONR program was important as the only funding agency program devoted to nonlinear dynamics as a field of physics, although other government agencies did fund nonlinear dynamics. In my program, initial thrusts were developing the mathematics of nonlinear equations, seeking physics topics described by nonlinear equations, looking for experimental verifications, and application to optimizing devices or creating new devices.

Advances in science can be surprising. An unexpected application of chaos was creating morphable computer chip logic gates. This type of gate is comprised of a single chaotic element that can switch (morph) its action among all the types of logic gates with a voltage change. Start with a quadratic map in a chaotic state that has a collection of unstable periodic orbits. A reset threshold voltage value can select values below and above it as a 0 and a 1. Any combination of inputs can produce any output by a selected voltage setting to produce any of the AND, OR, XOR, NAND, NOR, NOT gates. This was ONR-funded work by William Ditto and colleagues that is now using AI to select the configuration of gates to optimize an application. As an example, a circuit could, by morphing its logic gates, optimize signal to noise in a detector or improve pattern recognition.

If an ONR Program Manager has an idea and wants to start a new program, collaborate, or go somewhere in the world, ONR can provide the needed support. From day one, until I retired after over 40 years later, I managed a nonlinear science program. There was much more besides that. There were many activities and one could compete for special programs, including Small Business programs, Young Investigator program, and

Multidisciplinary University Research Initiative. Some of these provided the opportunity to work with expert colleagues, for example, we edited books on ocean waves, *Nonlinear Dynamics of Ocean Waves*, A. Brandt, S. Ramberg, and M. Shlesinger, and on ocean structures, *Stochastically Excited Nonlinear Ocean Structures* T. Swean and M. Shlesinger.

One example for starting a new program was inspired by supplying the military deployment at Mogadishu, Somalia. The port was within enemy fire range from militias, so cargo transfers were done at sea. The large cargo ships used cranes to transfer supplies to lighter vessels. The problem was that at Sea State 3 and above, the cargo ships rocked with the swells while the small ships responded to the chop causing the cranes' loads to swing and not be controlled. I started a program in crane control and discussed it with the Chief of Naval Operations during an ONR visit. I funded a small business that constructed a scaled-down wooden crane and was able to demonstrate crane control on a rocking platform. With control turned off, the load would swing chaotically. In practice, the limiting factor in crane control was the speed of the motor to implement the needed corrections to the crane movement. At least the method worked up to Sea State 3. Randy Tagg at the University of Colorado/Denver developed the theory for crane control and taught it in his physics class, where one of the students, Caleb Carr, founded a company, Vita Inclinata, and applied the dynamical control knowledge to helicopter lifting crane operation, preventing load rotation. Very important for lifting wounded personnel on a stretcher.

One year, our Director submitted to the Defense, Director Research and Engineering, the list of ONR MURI topics for approval. He had put in his own topic of the dolphin genome for clues on how dolphins can dive deeply without the bends. His topic was rejected, and early that morning, he sent an email asking if anyone had a replacement topic. This was strange because each department submits extra topics. Nevertheless, I quickly wrote up a topic on magnetometers, thinking about a solid-state device, a magnetic tunneling junction device. The ONR Director accepted this. Later that day, I got help from Kristl Hathaway, a magnetics expert from our Electronics Division, to rewrite the topic, improving all aspects. My topic was selected and proposals came in. Some people believe that topics are written with a person in mind and that a fair level playing field is absent. Well, the winner was a joint group from Berkeley and Princeton, where I did not know anyone, and the approach was novel to me: a vapor magnetometer. The idea is that an atom with a valence electron will have a Larmor precession in whatever magnetic field it is experiencing. Place a vapor of these atoms in a glass vial along with a diode laser fixed to the present Larmor frequency. If the magnetic field changes, then the laser becomes off resonance (detuned) by an amount related to the disturbing magnetic field. The sensitivity was astonishingly good, getting down to picotesla per root Hz. Naval interest included locating submarines. The group also had a spin exchange relaxation free (SERF) magnetometer at play in the femtotesla per root Hz, also with commercialization potential for medical brain and heart measurements. This program was a great success that arose from a level playing field and not my own limited ideas.

Double-Hatted

All through my 40 years at ONR, I was always a Program Officer managing a Nonlinear Physics core program. For many years, I had dual roles in higher management and teaching positions.

(SES) Physics division director

Joining ONR in September 1983 as a Scientific Officer was a full-time job. Note, later our designation became Program Officer. I did not appreciate the change in name and emphasis away from

"Science" to "Program". In 1986, in addition to my Scientific Officer position, I became the (acting) Physics Division Director, another full-time job. Having two jobs is what we called double-hatted. Every day, I did not leave until every travel request, travel voucher, and proposal approval submitted that day was approved. Apparently, that was not the norm in other divisions. I estimate that my work week became 60 hours. In part, I stayed late to wait for rush hour to wind down. I had an eighteen-mile return commute from Arlington, Virginia, to Rockville, Maryland, with several choke points on the way. Waiting until 7 pm made for a faster, less stressful ride home. Still, occasionally I needed to come in on the weekend to catch up on work that could only be done on the office computer. These were days before mobile phones and the Internet. In 1987, my Director position became permanent, and I was promoted to a Senior Executive Service (SES) level. This is the civilian equivalent to an Admiral in the Navy. I stayed double-hatted from 1986 to 1995 and then again from 2006 to 2010 under different positions.

The first ONR Division Director of Physics, in 1946, was a Johns Hopkins professor on loan. He went back to Johns Hopkins in 1948 with a grant from ONR as an appreciation for his contribution to ONR. Elliott Montroll, in 1948, was visiting Frederick Seitz at the Carnegie Institute of Technology (later becoming Carnegie Mellon University). Seitz was a well-known physicist and the author of the 1940 foundational book, *The Modern Theory of Solids*. He got a phone call asking for suggestions for a physicist to take over the Physics Division job at ONR. He turned to Montroll and asked if he was interested, and so Montroll became the second Physics Division Director, a position that he held from 1948 to 1950. In those days, ONR was in a Quonset hut on the National Mall, and Montroll bought a house in Chevy Chase, then considered to be far out from DC. Today, metropolitan DC goes many miles passed Chevy Chase and also in miles into Virginia.

Quoting Montroll about the early days of ONR. "Now, I had the pleasure of being head of the physics branch at Office of Naval Research at the time when money was first put into physics, which was very exciting, in the late forties, '48,'49. I found it very exciting, and people at ONR had a kind of personal independence that doesn't exist in institutions like NSF. We were more or less left alone to think of what kinds of projects, pick good people, to try to keep the work with some connection of topics that the Navy, that might be important to the Navy, but there were no review committees, no peer analysis of proposals. One was supposed to use good taste, and while it's now considered unfashionable to say this, but the Old Boy network was very good. And at that time, the network was very strong, because a community of good people was — everybody knew each other, and this was an exciting time to know everyone. One hardly had to be in the community very long. One could be 25 years old and already know all the important people in his field. Now one's lucky if he's 35, to get the same kind of responsibility".

In my time, it was still true that one was left alone to design and manage a research program. I did not know of or support an Old Boy network, although some older scientists would call (unsuccessfully), looking for a grant based on being an old boy. Each Program Officer in the ONR Physics Division had a physics topic: Atomic and Molecular, Plasma, Surface Physics, Lasers and Quantum Optics, Superconductivity, Acoustics, and my Nonlinear Physics. The Electronics Division had other physics topics. If someone with a great polymer physics topic applied to the quantum optics program, then that would fail. ONR neither had the funds nor the breadh of topics to fund all good proposals and topics. Go to DOE or NSF for particle physics or cosmology.

In 1970, Montroll became my PhD advisor at the University of Rochester, and in 1986, I followed his 1946 lead to become the

Physics Division Director. I took my position as the Physics Division Director seriously, believing I was part of a long tradition that would continue. I believed my leadership and accomplishments would be remembered, but that all came to a halt in 1995 in a reorganization when physics and chemistry were combined into a single division and moved to a Sea Warfare Department with barely a trace of physics history. ONR, in 1995, changed from Scientific Research Directorates overseeing Research Divisions to Warfare Departments. Now, I needed to compete for programs and funds based on their connection to Sea Warfare.

Around 1995, funding of DOD basic research (6.1 budget line item) for ONR, ARO, AFOSR, and DARPA, there was an internal DOD proposal being considered to combine all 6.1 funds into a single DOD agency. One suggestion was for ONR to manage all the Army, Navy, and Air Force 6.1 funds, and in exchange, ARO and AFOSR would receive more 6.2 and 6.3 funds. The Navy had 6.2 funds with the Office of Naval Technology (ONT) and 6.3 funds with the Office of Advanced Technology (OAT). Perhaps in a strategy to prevent the loss of 6.2 and 6.3 funds, ONT and OAT became part of ONR and the argument was made that ONR had now vertically integrated its 6.1 to 6.3 programs and could not combine this with Army or Air Force programs. That is when ONR switched to Warfare Departments.

Perhaps I wasn't the most aware SES in the building. We had a meeting of ONR's SES members led by our Executive Director, Fred Saalfeld. He started the meeting by asking if anyone doesn't know what TQM is. I raised my hand and everyone laughed. I was the only one with a raised hand. It turned out everyone was well acquainted with Total Quality Management. It was a business model of Edward Deming who applied his concepts to the rebuilding of Japan after WWII. He had fourteen points, one of which is "Break down barriers between departments. People in research, design, sales, and production must work as a team, to foresee problems of production and in use that may be encountered with the product or service". This meant that decisions from the top were passed by all levels of the company for input to avoid unintended consequences. Well, shortly after that meeting, an ONR reorganization was announced as fact, with no attempt to get input from all levels of the building. Screw Deming. As was remarked above, ONR gained the Navy 6.2 and 6.3 organizations. The Physics and Chemistry Divisions were combined and moved from the Research Directorate to the Engineering Directorate, while Computer Science was moved in the opposite direction to the Research side. I was offered to be head of joint Fluids and Physics group but turned that down. The newly formed Physics-Chemistry Division was now headed by a chemist. I stayed with Physics-Chemistry, but as a Chief Scientist with the Senior level designation of ST. In 1995, the Navy opened up all ST Chief Scientist positions for competition and I was fortunate enough to win one. It came with a four-page legal position description. Under President Carter, when the SES was created, ONR scooped up many of these positions, as people with super grades GS16-18 thought the SES would disappear with Carter, and it was prudent not to switch to it. Over time, ONR would lose many of the SES positions. Then, in March 2011, I received a letter directing that my ST position would be abolished in 2 years in March 2013. The ONR Executive Director, Walter Jones, came to my office to discuss the position abolishment letter. He asked how I was doing. I said, it's not a problem, I hope to stay at ONR, but if not possible, then I can find other jobs. I then said, I'm the happiest person at ONR. Walter said, that's refreshing, as everyone else has complaints. When March 2013 came around, I was changed then to what was called an NP-4 designation and was able to keep my ST salary and 720 hours end-of-year carryover annual leave. Non-SES-level position could only carry over 240 hours. The only actual change was that I was no longer in the ST/SES bonus pool but in the smaller Program Officer one. In these days of mass government firing, my ST abolishment letter was kinder and gentler, and I did stay happily at ONR until 2024, more or less with no changes.

(ST) Expeditionary maneuver warfare and combating terrorism division director

A new person became OSD's Defense Director of Research and Engineering. She held a competition for all Chief Scientist ST positions. Each position had a competition and an outside review; the Head of mine was Roger Dashen, a well-known theoretical physicist from UCSD, and I won the spot and was back to one hat, Chief Scientist for Nonlinear Physics. The one-hat situation lasted until 2007 when I became a Division Director again, this time in the new Marine Corps Warfare Department.

ONR created a new department in 2005 to manage the Marine Corps' 6.2 and 6.3 funds. This was combined with ONR's 6.1 funds for projects of Marine Corps interests. In 2007, I joined this new department as a Division Director, mostly for the 6.1 programs, and was once again double-hatted. With Navy Captain Mark Stoffel, I managed the Counter-Improvised Explosive Devices (CIED) program that had 75 projects. The workload included all program and budget reviews, and a good deal of traveling. Each Program Officer funded a representative at a Naval Warfare Center, and every annual review for each Program Officer would be held at a Naval Warfare Center. The Marine Corps programs focused on applied research, prototypes, and testing, not so much on basic research. Our Department worked closely with the Marine Corps Combat Development Command in Quantico. Thanks to Colonels Dodge and Kirby, twice I got out to Quantico to shoot M4 rifles and pistols. With the M4 illuminated gun site, I could hit some close-by targets, but nothing with the pistol. For the first rifle practice, I assumed an indoor range with paper targets that are pulled back for you to see your hits, like on TV shows. The first time, it was a cold winter day, and the targets were outside on a snow-filled field up to 800 yards away. I wasn't dressed warmly, and after some time, I was too cold to pull a trigger. I washed out playing Marine in under an hour. I was happy to warm up in the mess hall. We tried to levy humor into our full-day reviews. With one talk about a squad wearing haptic vests, so if, in the dead of night, the squad leader moved his hand from left to right across the vest, everyone felt on their vest the same sensation to get the order to move to the right. I asked if the leader scratched his ass, would the last man in the squad think it meant the Japs bombed Pearl Harbor? Anyway, it was funny at the time and a break from long hours of serious presentations. George Solhan, our Department Head, called for a recess until the laughter receded.

Kinnear Chair of Physics, United States Naval Academy

ONR allowed me to consult on my own time (using my earned vacation time). I worked with John Bendler at General Electric in Schenectady, NY, for 10 years with quarterly visits on topics of plastics and the glass transition. With Joseph Klafter at Exxon, we worked on statistical aspects of nonlinear dynamical systems. Details of some of these works are in my book, *An Unbounded Experience in Random Walks with Applications* (World Scientific Publishers). Note that the cover figure is my calculation of the Zaslavsky map for a kicked oscillator, modeling a periodically kicked charged particle in a magnetic field. For an ensemble of initial conditions, a stochastic separatrix web forms. An implication from this model is that it will be difficult to prevent nuclei from escaping in a fusion chamber, as the stochastic web affords escape paths. I met George Zaslavsky when he left the Soviet Union to come to the US. He had been following my work in Russia and contacted me when he came to Maryland. We became fast friends and co-authored (with Yossi Klafter) a paper for *Nature, Strange Kinetics*, about trajectories in nonlinear Hamiltonian systems. Dissipative dynamics can have strange attractors, so we named strange kinetics for dissipation-free systems. Our paper has over 1,000 citations. George moved to the Courant Institute in Manhattan, so I had to stop working with him as I began to fund his work.

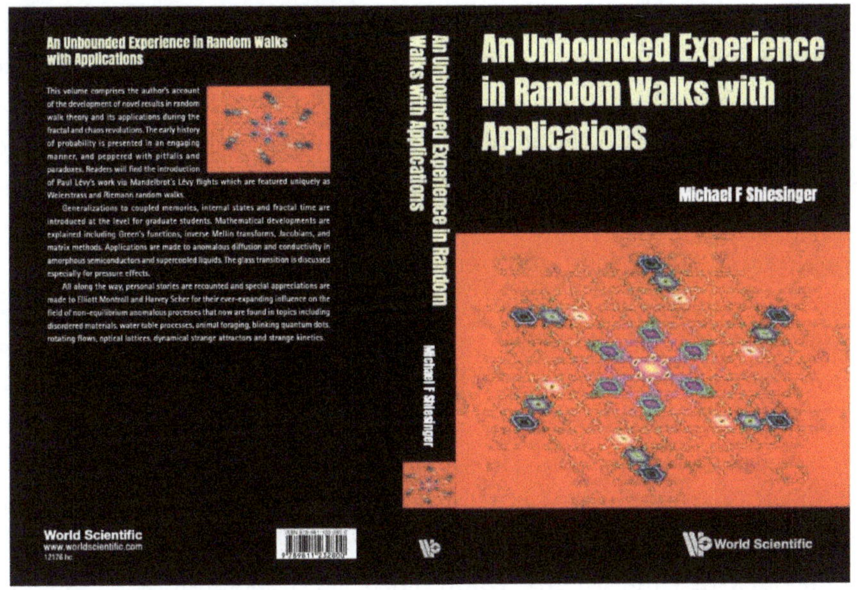

We were fortunate that two scientists at the USNA, in Annapolis, were following my work with John Bendler; they were John Fontanella and Mary Wintersgill. They had an exciting lab with many experimental results on polymer electrical properties under various conditions, including pressure effects with application to solid state batteries. We joined forces and made progress in understanding the properties of polymers, some are detailed in my *Unbounded Experience* book. In 2008, James Kinnear, a former USNA graduate and a former President and CEO of Texaco, endowed a Physics chair at the USNA. John Fontanella recommended me for the position. I was well known to the Provost, Admiral William Miller, as he was a former Chief of Naval Research, and he called the current CNR, Admiral William Landay, to ask for my reassignment to the USNA. I held the inaugural Kinnear Chair for 2 years and was again double-hatted, 2008–2010, still also working as an ONR Program Officer. I taught the second-year midshipman physics course (mechanics, electricity, magnetism, waves, fluids, sound, and light) and the solid-state physics course for seniors. For the introductory course, there was a curriculum to follow, lecture by lecture. I could have done better. When covering a topic, I would see some heads nodding in agreement, so I would move on to the next topic. I didn't see the heads not nodding in affirmation and was leaving some students behind. The final was for all students, the whole brigade of over 1,000. I had to move on to cover all topics for the final exam. As there are no graduate students, the professor needs to teach the labs, set and grade homework, and create and proctor exams. It is a 45-mile drive from my home to the Academy, and I did this 4 days a week, then 1 day a week at ONR. An advantage is that I did go swimming every day at a USNA pool, then lunch at the Drydock or the Officer's Club, then back in time for my 13:00 class. Here is the letter I sent to James Kinnear and his reply to me.

DEPARTMENT OF THE NAVY
UNITED STATES NAVAL ACADEMY
ANNAPOLIS, MARYLAND 21402-5000

Mr. James W. Kinnear III   August 6, 2009
149 Taconic Rd
Greenwich, Connecticut 06831-3113
Dear Mr. Kinnear,

I am honored to have been selected as the James W. and Mary T. Kinnear Chair in the Physical Sciences at the USNA and to have this opportunity to write to you. My two year appointment began on August 1, 2008 and I have spent the last year teaching electricity and magnetism, performing research, publishing in scientific journals, lecturing and as a divisional associate editor for the Physical Review Letters, the most prestigious physics journal.

Some background about myself. I have been with the Navy, through the Office of Naval Research (ONR) for 26 years. At ONR I have been the Director of the Physics Division, the Research Director of the Expeditionary Maneuver Warfare Department (USMC programs) and am presently in the role of a Chief Scientist. During my two year detail to the USNA, I work 80% of my time at the USNA and 20% at ONR. My travels are paid by ONR. I have published over 200 scientific papers, presented over 350 lectures worldwide, hold patents on peptide drug design, received the Senior Executive Services' Presidential Rank Award for Senior Professionals, a Navy Superior Civilian Service Award, and the ONR Saalfeld Distinguished Lifetime Achievement Award. In addition, I was the Michelson Lecturer at the Academy in 1992, the Regents Lecturer at UCSD, and I received the Medal of Valor from the city of Rockville, Maryland for saving the lives of two people.

The Kinnear Professorship has afforded me the opportunity to work with USNA Physics professors on the topic of the glass transition. This involves the physics of how a liquid can turn into a glass (and not a crystal) upon rapid cooling. Applications include, for polymeric materials, ion-doped polymeric batteries and slowing the aging of bullet proof polycarbonate glass. Our work (theory and experiment) involves a host of kinetic and thermodynamics questions pertaining to dielectric relaxation, conductivity, and viscosity as a function of temperature, pressure, rate of cooling and doping with ions. We have made steady progress in our research and have published several papers over the past year, acknowledging the Kinnear Chair, and have several more publications in preparation. I've presented nine lectures this year, including talks at Princeton, USC, UC Irvine, and the Washington Philosophical Society.

My most rewarding experience has been teaching and working with the Midshipman. Although not all of my students are interested in Physics, they all have demonstrated a strong work ethic, sense of fun and camaraderie. It was a joy to be with them and I thank you for making this experience possible. I look forward to teaching classical mechanics in the Fall semester.

Very respectfully,
Dr. Michael Shlesinger
shlesing@usna.edu / 410-293-6614
Physics Dept. USNA Annapolis MD 21402
James W. and Mary T. Kinnear Chair in the Physical Sciences
ONR Chief Scientist

JAMES W. KINNEAR
TWO STAMFORD PLAZA
281 TRESSER BOULEVARD, SUITE 1500
STAMFORD, CT 06901

PHONE: (203) 327-7858
FAX: (203) 327-7861

August 18, 2009

Dr. Michael Shlesinger
Department of the Navy
United States Naval Academy
Physics Department
121 Blake Road
Annapolis, MD 21402

Dear Dr. Shlesinger:

Thank you for your very kind and inspiring letter. Perhaps you saw the piece in the USNA fund raising bulletin in which I said that you are providing just the sort of fine academic leadership that my wife, Mary, and I had in mind when we funded the chair.

You and I have more in common that just the love of physics! I was also decorated for saving a man's life.

I come to the Academy quite often and will make sure that I call on you next time. I want to know more about your fascinating research.

Thank you for taking the time to write. I am proud to have our names associated with you.

With kindest regards, I am,

Sincerely,

JWK:dal
Enclosure

NRL and the Naval Warfare Centers

ONR supported with 6.1 funds independent research at the Naval Warfare Centers. There was discussion that the funds were better spent elsewhere, such as at universities. The Navy uses nine Technical Readiness Levels (TRL). The Warfare Centers were to move work at TRL 3 (Analytical and Experimental Critical Function and/or Characteristic Proof of concept) to TRL 6

(System/subsystem model or prototype demonstration in a relevant environment), and not to be doing 6.1 basic research. Naval work eventually transitions to and then buys from industry. As basic research was being questioned at Navy labs, I sent to John Deutch, then an Under Secretary of Defense, the 1993 volume of the Naval Research Reviews that I edited, focusing on my nonlinear physics program, which includes articles from independent research from NRL, NSWC, and SPAWAR (now NIWC).

The Naval Research Reviews articles were *Controlling Chaos* by Mark Spano and William Ditto from NSWC, *Communicating with Chaos* by Thomas Carroll and Louis Pecora from NRL, *Nonlinear Resonance* by Adi Bulsara, John Douglas, and Frank Moss with Bulsara from NIWC/Pacific, *Noisy Chaos* by Robert Cawley and Guan-Hsong Hsu from NSWC, and *Attractors of Iterated Systems Application to Image Compression* from E. W. Jacobs and R. D. Boss from NIWC/Pacific, and I wrote about *Applications of Chaos and Fractals* from ONR.

The response from Deutsch was that the Navy was getting its money's worth supporting basic research at the Warfare Centers, and my work was appreciated. My Department Head thought it was unusual for me to be interacting directly with a high level in the Pentagon. Till today, some of the programs I supported are having an impact recognized within the Pentagon.

Here are some of the Navy lab programs supported within ONR's Nonlinear Physics program.

Chaos control (NSWC)

At NSWC/White Oak, an experiment on the buckling of a magnetoelastic metal ribbon whose Young's modulus can be decreased in a magnetic field. When placed in a DC plus a 1 Hz AC magnetic field, the ribbon buckles in a chaotic manner. In the chaotic dynamical state, a return map of successive positions (a delay coordinate embedding) has the points fall on a single hump curve. This allows prediction from one buckling to the next and allows changes in the magnetic field to pick out and control unstable orbits, the first demonstration of chaos control achieved by Mark Spano and Willian Ditto. A practical ONR-funded example from Raj Roy, then at Georgia Tech, is when a blue-green laser is pump power driven to a too high intensity, then the optical production becomes chaotic. Applying chaos control can generate a higher intensity laser for underwater communications.

The University of Maryland group of Edward Ott, Celso Grebogi, and James Yorke (OGY) developed the theory for chaos control, and they strongly interacted with the NSWC group. OGY and students probably contributed the most world-wide advances in nonlinear dynamics theory as recounted in Ott's textbook *Chaos in Dynamical Systems*.

Chaos synchronization (NRL)

NRL is ONR's corporate lab and not a Warfare Center. It is a working capital fund lab meaning it must generate its own funding. While ONR provides around 20% of its budget and then it needs to raise the other 80%. The work is so important and high quality that each year the additional funds are forthcoming. One project funded by my Nonlinear Physics program is a classic work on the synchronization of chaotic systems, by Lou Pecora and Tom Carroll, for example, between a transmitter and a receiver. For secure communications, a signal could be embedded in a chaotic background, and only the synchronized receiver can subtract out the chaos and read the signal. The chaos can be generated as a high-dimensional signal from a circuit comprised of several variables and parameters. The receiver would be a stable subsystem missing part of the transmitter circuit, but receiving a signal would be equivalent to replacing the missing transmitter circuit part. Chaotic synchronization has been demonstrated. Consider a cell phone that encrypts a conversation that only the specialized receiver can decrypt.

Fluxgate magnetometers (NIWC/Pacific)

This type of magnetometer has a magnetic core and a driving and sensor wire coils. An AC current applied to the driving coil induces a magnetized, unmagnetized two-state field cycle in the core, inducing a voltage in the sensing coil.

A magnetic field is detected when the input current and output voltage signal no longer match described by the residence time asymmetry between the two states. This is modeled as an overdamped bistable nonlinear driven system in the potential:

$$V(x) = -ax^2 + bx^4$$

The middle figure is the non-driven case, and the system rests in one of the two wells. Sinusoidal driving raises the potential, allowing the system to switch between the two wells. For the magnetic core comprising many spins, the average spin cancels out $\langle S \rangle = 0$. When an AC magnetic field is applied, a spin will oscillate between spin up $= +1$ and spin down $= -1$. From In and Palacios' book, *Symmetry in Complex Network Systems*, the average magnetic field $\langle h_i \rangle$ at spin S_i with contribution from all the other spins is modeled as

$$\langle h_i \rangle = \sum_j w_{ij}\langle S_i \rangle + h$$

where h is the external magnetic field and all the w's are set equal to $1/N$ for an N-spin system. We choose a function that ranges between -1 and 1:

$$\langle S_i \rangle = \tanh\left(\langle h_i \rangle / kT\right)$$

substituting for $\langle h_i \rangle$ and summing over all i we have,

$$\langle S \rangle = \tanh\left(\left(\langle S \rangle + h\right)/kT\right)$$

where $\tanh$ is a function that spans from -1 to 1. In differential form for continuous time,

$$\tau \frac{d\langle S \rangle}{dt} = -\langle S \rangle + \tanh\left(\left(\langle S \rangle + h\right)/kT\right)$$

The point now is for h, our applied AC magnetic field, to include a DC component that one is trying to detect. This target component will cause an asymmetry measured by the residence time between each of the two bistable potential wells. Adi Bulsara and colleagues made two major advances for the fluxgate magnetometer. First, Bulsara discovered that adding a precise level of noise improved a weak DC signal detection. This is called stochastic resonance. Second, they coupled several magnetometers, for example, for three:

$$\frac{dS_i}{dt} = -S_i + \tanh\left(3\left(S_i + \lambda S_{i+1}\right)\right)$$

for $i = 1, 2, 3$ and with $i = 3$ cyclically connected to λS_1 to improve the detection of a weak DC signal.

Channelizer: Dynamical analog to digital converter (NIWC/Pacific)

Push a pendulum with any amount of force, and if no friction, it will oscillate forever. The oscillating LC circuit was noted earlier. Another electronic example is the van der Pol oscillator with slow and fast sections of the oscillation cycle. There is the Hopf bifurcation where an increase in a parameter moves a trajectory from a stable point to limit cycle oscillations. The Lotka–Volterra coupled nonlinear equations produce oscillations as a model of predator–prey interactions. There is a world of models for nonlinear oscillations. The NIWC team introduced one based on symmetry for applications for an analog-to-digital converter and for precise timekeeping.

The Naval Information Warfare Center (NIWC) at Point Loma in San Diego has made several advances in nonlinear physics. At NIWC, Visarath In developed a chip with a progressive sequence of coupled nonlinear oscillators. The elements are the overdamped bistable oscillators described for the fluxgate magnetometer. Three of these are coupled, and complex nonlinear behavior ensues depending on the coupling configuration and strength. One of the resulting behaviors of this three-coupled element system is for the system only to oscillate when driven at a fixed resonant frequency. An array of these systems is configured, so each is resonant with a different frequency to capture a spectrum of frequencies. This performs a spectral decomposition when picking up an analog signal. The channelizer performs a dynamic analog-to-digital conversion.

What are the military implications? Place this chip on a number of small aircraft drones not radar visible. When interrogated by radar, it can convert the radar into a digital signal, have a system to amplify it, and transmit it back. The adversary now thinks its radar hit a target. If the transmitted signal is delayed, it gives a false distance response. The idea is to confuse an adversary of where your air forces are by presenting false radar responses from a bunch of drones. See *Symmetry in Complex Network Systems* by Visarath In and Antonio Palacios for more examples from the NIWC team, including the nonlinear dynamics of coupled fluxgate magnetometers, precision timing with crystal oscillators in GPS denied areas, spin torque nano-oscillators for high power microwave beams, and coupled energy harvesters.

The Navy labs were always the bulwark of my ONR program with relevant applications. The scientists could have been at any top university; I'm grateful that they were supporting Naval labs with innovative state-of-the-art research.

Agent-based software (CNA)

The Center for Naval Analyses is an independent, non-profit research and analysis organization, but I've included it in this section along with the Naval Warfare Centers.

At University of Rochester, Freeman Dyson, around 1970, gave a lecture that confused people, including myself. It was about self-replicating entities that had some amount of free will. People asked if these were robots, and the answer was no. The entities had some kind of computational power. I met Lt. General Paul Van Riper in Secretary of Defense William Perry's office. Not long afterwards, General Van Riper introduced me to Andy Illachinski from the Center for Naval Analysis. Andy had developed an agent-based software to model land-based combat for the Marine Corps. The computer agents could cooperate and were each one was governed by a set of rules and attributes. A particular agent could be programmed to favor defense or offense, and to seek a degree of cooperation. The variety of agents could exhibit self-organization and realistic emergent behavior consistent with Marine Corps tactics. Although the simulations were for land warfare, the program was later generalized and could model scenarios from helicopter reconnaissance to missile defense. Through ONR, I funded Andy's work. It progressed rapidly, some of which can be found in his book, *Artificial War*. I suggested to a JASON member to invite Andy to their Washington meeting. As Andy began his lecture at the JASON meeting, I saw Dyson perk up with rapt attention. It finally dawned on me that in 1970, Dyson's lecture was ahead of its time and was on agent-based software. As an aside, during a lunch break, Dyson mentioned that he enjoyed my 1996 Physics Today article *Beyond Brownian Motion* (BBM) discussing various fractal properties. Dyson has a famous work on the eigenvalues of random matrices whose elements execute Brownian motion. In a discussion with Mandelbrot, BBM are his initials, and he hadn't realized this Easter Egg in the title of the Physics Today article.

Meetings

With NRL and NIWC leadership, ONR sponsored nonlinear dynamical conferences. One, in 2008, at Amelia Island in Florida coincided with my 60th birthday. Here's a list of my colleagues/organizers/friends and a photo from the meeting of three nonlinear dynamics stars (left to right), William Ditto (chaos computing), Lou Pecora (chaos synchronization), Mark Spano (chaos control), and then myself.

Amelia Island Organizers

Derek Abbott, Dept. of Electrical Engineering, Univ. of Adelaide
Jay Banavar, Dept. of Physics, Pennsylvania State Univ.
Yuri Braiman, Oak Ridge National Laboratory.
Adi Bulsara, SPAWAR Systems Center
Tom Carroll, Naval Research Laboratory
Jim Crutchfield, Complexity Sciences Center, Univ. of California, Davis
Mingzhou Ding, Dept. of Biomedical Engineering, Univ. of Florida
William Ditto, Dept. of Biomedical Engineering, Univ. of Florida
Fereydoon Family, Dept. of Physics, Emory Univ.
Mike Gabbay, Information Systems Laboratories, Inc.
Dan Gauthier, Dept. of Physics, Duke Univ.
Bruce Gluckman, Center for Neural Engineering, Penn State
Frank Gordon, SPAWAR Systems Center
Visarath In, SPAWAR Systems Center
Ying-Cheng Lai, Electrical Engineering Dept., Arizona State Univ.
Gloria Lubkin, Physics Today
Arnie Mandell, Cielo Institute
Ed Ott, Physics Department, Univ. of Maryland
Louis Pecora, US Naval Research Laboratory
Jeff Rogers, Dynamical Systems, California Institute of Technology
Karen Selz, Cielo Institute
Ken Showalter, Department of Chemistry, West Virginia Univ.
Tom Solomon, Dept. of Physics, Bucknell Univ.
Mark Spano, NSWC, Carderock Laboratory
Harry Swinney, Dept. of Physics, Univ. of Texas
Randy Tagg, Physics Dept., Georgia Institute of Technology
Bruce West, Army Research Office
Kurt Wiesenfeld, Physics Dept., Georgia Institute of Technology
John Zimmerman, Office of Naval Research, Global

Adi Bulsara and Visarath In at NIWC/Pacific at Point Loma in San Diego created the ONR-sponsored International Conference on Applied Nonlinear Dynamics (ICAND) to focus on applications held in Catania, Kauai, Lake Louise, Seattle, Denver, and Maui. The Hawaii locations were to attract scientists from Asia.

One famous meeting was held in Lanzhou, China. Grebogi and Ott had an exceptional PhD student at the University of Maryland, Ying-Cheng Lai. See his book with Tamas Tel, *Transient Chaos*. His research includes topics on chaotic scattering, noise-induced chaos, and relativistic quantum chaos. Lai, now a chaired professor at Arizona State University, had an exceptional PhD student, Liang Huang. Grebogi, now at the University of Aberdeen, Scotland, joined with Lai and Huang to create the Aberdeen-Lanzhou-Tempe Joint Research Center for Computation and Complexity. In the photo, I'm in the back row next to Liming Salvino on my left, with her representing ONR Global. Huang is third from the left in the back row. Grebogi and Lai, on his left, are signing for Aberdeen and Tempe to establish the Research Center. Grebogi and I gave the keynote addresses. My talk was on the history of probability, so as not to run afoul crossing any security questions about discussing Naval research topics that might be classified.

Poster advertising the Grebogi and Shlesinger lectures at the inaugural Lanzhou Center for Complexity Research.

Programs

At ONR, my nonlinear science program focused on instability, how to predict it, cause or avoid it, and exploit it.

Much of good engineering has focused on stability and linear response, where a small disturbance leads to a small change. Developments continued on instability in nonlinear dynamics. Just consider synchronization through nonlinear forces. Early advances in nonlinear physics start with Huygens, in 1655, with observing the dynamical synchronization of wall-mounted pendulum clocks locked in amplitude and anti-phase, interacting through wall vibrations. Lord Rayleigh, in his 1877 *Theory of Sound*, described acoustic synchronization in pipes, and in the 1920s Van der Pol studied synchronization of nonlinear electrical oscillations in triode vacuum tubes for radio communications. Kuramoto, in 1975, modeled a large array of globally coupled identical nonlinear oscillators, finding an order parameter that governs a transition to a synchronized state.

Continued research into synchronization as a promising scientific research area includes non-identical oscillators synchronization using large bandwidth chaotic oscillators to expand the possibility of frequency, amplitude, and phase locking. Synchronization phenomenon is a prevalent instance of cooperative behavior, manifesting itself in a wide spectrum of real-world systems. In technological systems, synchronization is required for an array of lasers to emit high intensity radiation, for arrays of superconducting Josephson junctions to emit high-intensity terahertz radiation, for the operation of coupled power generators in AC electrical power grids, and many other applications. Another example is in the field of spintronics, where electron spin is exploited. The physics of synchronizing many spins is sought to lead to the design and fabrication of smaller, faster, more sensitive, more energy-efficient nano-devices. Today's state of the art has work that demonstrated disordered homogeneous arrays resulting in synchronization and chaos elimination in the arrays of nonlinear oscillators. Following that work, additional evidence was presented on the effects of disorder in eliminating chaos and enhancing synchronization in dynamical systems or inducing dynamical order through resonance-like phenomena in driven system. More recently, the following was demonstrated: (i) complete synchronization is common in a wide range of non-identical oscillators even when oscillator heterogeneity is large and (ii) oscillator heterogeneity can have a constructive role in stabilizing otherwise unstable states of complete synchronization. In laser networks, connecting laser cavities through a combination of heterogeneous dissipative and dispersive coupling mediated through a scattering element which inevitably causes radiation leakage opens the door to effectively control a self-organized frequency synchronization and select a frequency mechanism of its emergence. In 1665, who could say where Huygens' clocks could lead? The same is true for today's science and predictions of the future.

I helped create a MURI topic on nonlinear synchronization with bright prospects for applications.

Returning to the question of funding research projects, what is the payoff? That can be asked from the beginning of math and physics. Basic research for its own purpose and enjoyment in many instances leads to large and sometimes surprising payoffs. But it all starts with basic research in science and mathematics. So, thanks to those geniuses who over thousands of years pushed knowledge and ideas forward without knowing it would lead to lasers, computers, jet engines, and all our modern wonders and worries.

Here's a list of my programs: Each program had to be designed, proposed, and competed for funds. Every day was a busy day.

- **Nonlinear Physics** (Core 1983–2024)
- **Nonlinear Phenomena** (Accelerated Research Initiative, 1986–1990)
- **Fractal Clusters** (Accelerated Research Initiative, 1986–1992)
- **The Glass Transition** (Core Enhancement, 1987)
- **Fluid Chaos and Turbulence** (DARPA University Research Initiative, 1986–1991)
- **Fractal Image Compression** (Small Business Innovative Research, Phase I 1990, Phase II 1991–1992)
- **Nonlinear Dynamics of Ocean Waves** (Accelerated Research Initiative, 1992–1996)
- **Nonlinear Signal Processing** (Small Business Innovative Research, Phase I 1992, Phase II 1993–1994)
- **Nonlinear Dynamic Control of Lasers** (Small Business Innovative Research, Phase I 1993, Phase II 1994–1995)
- **Nonlinear Extremal Techniques: Dynamics and Control of Randomly Forced Nonlinear Oscillators** (University Research Initiative, 1993–1995)
- **Dynamical Neural Systems** (Accelerated Research Initiative, 1995–1999)
- **Stochastic Resonators** (Small Business Innovative Research, Phase I 1994)
- **Chaos Control of Shipborne Cranes** (Small Business Innovative Research, Phase I & II, 1995–2000)
- **Nonlinear Control** (Multidisciplinary University Research Initiative, 1996–2000)
- **Nonlinear Devices** (CNR enhancement 2000–2002)

- **Thermal Combustor Nonlinear Control** (Small Business Innovative Research, Phase I & II, 2004–2006)
- **Magnetic Detection Science & Technology** (Multidisciplinary University Research Initiative, 2005–2009)
- **Nonlinear Dynamics for Novel Devices** (Multidisciplinary University Research Initiative, 2007–2011)
- **Counter-Improvised Explosive Devices** (2006–2009, 2012–2015)
- **Morphable Dynamic Information Processing** (Multidisciplinary University Research Initiative, 2012–2016)
- **Computing with Chaos** (STTR, 2012 Phase I & II)
- **Sound and Electromagnetic Interacting Waves** (Multidisciplinary University Research Initiative, 2010–2014, inherited in 2012)
- **Random Lasers, Nano-Spasers, and Optical Rogue Waves** (Multidisciplinary University Research Initiative for 2013–2018)
- **Cognitive Neuroscience of Memory Consolidation Across Sleep Stages and Efficient Learning** (Multidisciplinary University Research Initiative for 2013–2017, code 34 lead)
- **Multimodal Biosensing** (STTR, Phase I 2014, Phase II 2015)
- **Mathematical Methods for Deep Learning** (MURI 2020–2024 code 311 lead)

I would have liked to do more, but only so much is possible. For example, John Hopfield came to ONR to present a seminar on neural networks. Right afterwards, I would have liked to start a neural network program but didn't have the time. Our Electronics Division started a great neural networks program, so ONR was well covered on the topic without me. Decades later,

I did get involved with deep learning AI, which grew out of neural networks.

Briefings

I could find myself at the Pentagon briefing, Richard Danzig, the Under Secretary of the Navy and his three-star group about chaos. Note that at the Pentagon, they use a seating chart. I'm at the podium, and a Major General is changing my slides in the pre-PowerPoint days.

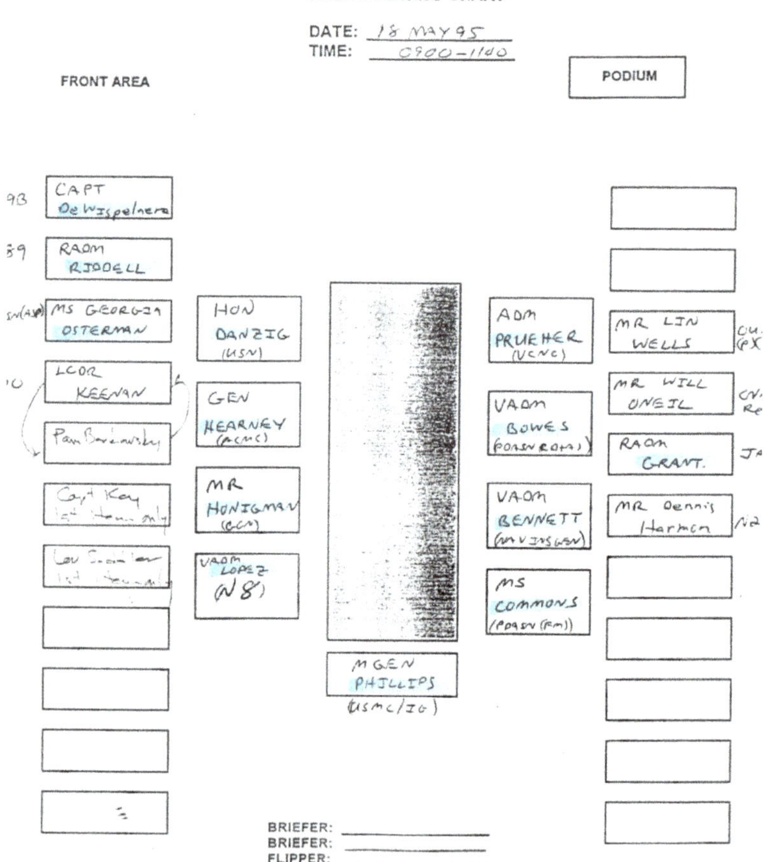

Then, I'm back at the Pentagon briefing Bill Perry (Secretary of Defense), Anita Jones, a computer scientist (Director, Defense Science and Engineering), and Lt. General Paul van Riper (Marine Corps Combat Development Command). Dr. Perry had a PhD in mathematics, and each week, when possible, he invited people to discuss science and math topics. This was a pleasant break from the serious responsibilities of being the SECDEF. It went well, and General van Riper and I met later to discuss agent-based software for warfare simulation, and I began to support Andrew Ilachinski at the Center for Naval Analyses on this topic. See Andy's book, *Artificial Warfare*. The thank you I received from Dr. Jones after the SECDEF briefing read, "I want to thank you for the excellent briefing that you provided to the Secretary of Defense today. From my observation, Dr. Perry enjoyed your presentation on Nonlinear Dynamics. Your experience and hard work certainly has earned our respect". I briefed Vice President Al Gore's staff on nonlinear dynamics. They asked how many jobs it could create. I did not have an answer.

14

Saalfeld and Weisskopf Awards

Fred Saalfeld was the legendary Executive Director of ONR. He spent around 20 years at the Naval Research Lab where he was the Head of the Chemistry Division. At NRL, he created a spectrometer that is used to test the air on submarines. He moved to ONR for the next 20 years, first as the Research Director and then as the Executive Director overseeing all aspects of ONR including research, financial, contracts, NRL and Naval Warfare Center interactions, ONR Global, Congressional Liaison, building maintenance, and a few others. I don't think I ever came to the office before him or left after him. He was on top of every aspect of ONR. Fred sent me to the Pentagon to brief and make connections with several Admirals. I especially enjoyed the interaction with Adm. Craig Dorman and visited with him later when he became the Director of the Woods Hole Oceanographic Institute. It was great to work with Fred. When he retired from ONR, the Frederick E. Saalfeld Award was created to honor

Outstanding Lifetime Achievement in Science. The first award went to a Harvard physicist. I received the second award in 2006.

Marlan Scully at Texas A&M University's Institute for Quantum Science and Engineering leads and mentors a combined theory and experimental group. He portrays what a university is supposed to be about: research, scholarship, teaching, community, support, and a growing international family. Marlan created the Physics of Quantum Electronics (PQE) conference in 1968, and it continues to go strong with the next meeting at Snowbird on January 5–9, 2026. In 2024, through PQE, Marlan created the Victor Weisskopf Medal, and I am the recipient of the inaugural award.

15

Schrödinger's Dogs and Yellow Pearls

Schrödinger discussed about a cat being in a quantum super-position of alive and dead states to address the quantum measurement question. This is known as Schrödinger's cat question. Apparently, Einstein had discussed this question in terms of an explosive being stable or exploded quantum state. In any event, this is a story about dogs while I was writing a book review of Schrödinger's biography.

The sound was otherworldly. I was in my library typing away on my Apple IIsi with a great little word processor called Multi-Scribe. I was writing a book review of Moore's biography of Schrödinger. I knew nothing personal about the man, and the book was fascinating. After surviving World War I, as an Austrian soldier on the Italian front, he decided to live it up. And he did. Somehow, in the midst of this, he managed to discover the wave function equation of quantum mechanics while on a holiday with his mistress. Permeating through all this were growls and screams. What? Yes, that's exactly what I thought, what on earth is going on?

The growls were easy to identify. They must be from those strange dogs that moved in across the street, with their owners. These dogs acted as if they owned the neighborhood, and their owners seemed proud of that. I later learned that the dogs roamed free, unleashed at night, startling people. I just knew the dogs were a constant nuisance, always barking and challenging. It was their frenzied barks I heard now this Saturday morning in 1990. What had they done? Chased a cat up a tree, or chased a squirrel? Whatever it was, it disturbed my writing. I got up and opened the front door.

The scene looked like a National Geographic special of wild dogs in some foreign land pulling down prey. Two of the dogs had effectively seized a woman with their teeth and prevented her from moving. The third, larger dog was preparing to take the woman down while she was holding her baby up with one arm. The dogs, a mother and two offspring, were hunting. The woman, I later discovered, had a hearing impairment, so her screams were somewhat distorted and confusing to me when I was sitting at my computer. But all was clear now.

Once outside, viewing this surreal scene, without thought, I darted over, took the child, and yelled at and pushed the dogs. They ignored me. No one else was in view, and the street was empty. Not knowing which house might be open, I headed back to my house although it was not the closest. Pulling the mother and holding the child, the dogs attacked again, but not me, only the woman. Perhaps her hearing aid had a frequency that set off the dogs? We struggled with blood, flesh, and screams, and I fell down still holding the child. In this short time, I was breathing heavily and losing the battle. Just then, my wife came out running and took the child. Now with two arms free, I picked up the woman and headed for my front door. We all made it back, and I called 911, but initially, I was too out of breath to speak.

The woman needed hospitalization and is well today, and the child was unscathed.

After an ambulance and the police came, I went outside and saw strings of yellow pearls. What was that? It was fat ripped out of the body, and that was the way it congealed.

The dogs, I later learned, were Rottweilers, and I've since been attuned to news reports of dog attacks, some fatal. The whole episode only took a few minutes. After which I returned to my computer and finished the Schrödinger book review (*J. Stat. Phys. 62*, 877–878 (1992)). The city of Rockville gave me their highest award, the Medal of Valor Award at their annual police awards dinner. I also received official citations from the Maryland Senate and House of Delegates.

For my 60th birthday, a colleague in Australia created a Wikipedia page for me, and some parts are even accurate. Later, my students at the USNA added a note about my receiving the award for saving the lives from the Rottweiler attack. Then, someone removed that added note from Wikipedia with the explanation that they couldn't find verification on Google. Strange world we live in.

Mayor and Council – Rockville, Maryland

Public Safety Award Certificate

Dr. Michael Shlesinger

Medal of Valor
June 11, 1991

Douglas M. Duncan
Mayor

Kay Mitchel
Chairman

T.N. Treschuk
Chief, Rockville City Police

16

Random Stories and Thoughts: Dumb and Otherwise

Dumb things I overheard in high school and college.

In high school, one guy said he wanted to work in the Post Office delivering mail because you can finish your route early and go bowling every day. He thought bowling would be popular in the future and that government jobs were easy. Nothing was further from the truth.

In college, where marijuana was plentiful, one guy said, when he graduates, he'll rent a boat in Tijuana, load it with marijuana, and sail back to the US. He was not a science major.

In college at Stony Brook, there was a protest against the Vietnam War with demonstrators facing off against a police line. I overheard students from the radical SDS saying, let's stand behind the crowd and throw rocks at the police. See if we can get them to overreact and harm students. It will only increase our cause. A nasty attempt to cause a riot and harm students. Not much different from terrorist group thinking.

The bombing of Cambodia caused further riots, so much so that the university President called off final exams so students could demonstrate for peace. Instead, most went to Long Island's popular Jones Beach. By the way, the Physics Department did hold final exams, which I was happy to take.

Physics stories.

It was said that nonlinear dynamics and chaos theory would have been done 100 years ago if computers were invented and quantum mechanics was not.

Funny stuff that passes through ONR.

A Russian at a cocktail party was bragging that American submarines could be found using the Aharonov–Bohm effect. What's that? That threat was brought to ONR to Adm. Gaffney. The Admiral brought the topic to me. Going back to the Maxwell equations, since there are no magnetic monopoles,

$$\nabla \cdot B = 0, \quad B = \nabla \times A$$

Mathematically, this means that B can be written as the curl of another vector, which has been called A, the vector potential. While it seems that only B is real and A is just a mathematical expression, Aharonov and Bohm proposed an experiment to show that A can be measured in its own right. In fact, A is prominent in the Josephson effect for tunneling electron Cooper pairs for measuring magnetic flux Φ through a surface in a Superconducting Quantum Interference Device (SQUID):

$$\Phi = e^{i \oint A \cdot ds}$$

So, the joke was naming a well-known type of magnetometer by a physics effect. I explained this to a relevant submarine group. But the joke sent a ripple through the intelligence community, which ONR was able to erase their concern.

Some work I funded may have had relevance to submarine detection. One of the Naval Warfare Centers presented some results to the HASC in Congress, which led to my phone ringing. It was Congressman Norm Dicks the Chairman of HASC. I said, I'm not allowed to talk with you, I would need to go through my chain-of-command to speak with Congress. He said, the hell you can't. So, we had a good conversation. Afterwards, I went to our Executive Director, Fred Saalfeld, to report my discussion with the HASC. He was not happy and sent me to confess to the CNR, Admiral Gaffney, to get reprimanded. The Admiral was more than pleased as this created an opportunity for him to meet with Congressman Dicks. The result was that I ran a classified meeting at NIWC/Pacific and the ONR-funded research continued.

I was called to a meeting with the ONR Executive Director and a retired Army Colonel. The Colonel knew for a fact that an alien spaceship had been observing the Vietnam War and was interfering with our satellites at geostationary points. He asked what ONR is going to do about it. Our Director said we fund research, and that alien spacecraft are beyond our charter. The colonel left, saying he had to catch a flight to Afghanistan. He sounded like a character from Dr. Strangelove.

A Russian scientist moved to the US and said he had a signal processing method to find American submarines. I was told an exercise was set up. A portion of the ocean was designated as having a tic-tac-toe pattern, and a submarine was directed to one of the nine boxes. On the first try, the Russian picked the correct box. Very unsettling. On all the other attempts, he got the wrong answer, so his supposed method was worse than guessing. It turned out he was ill and came to the US to get medical care and hoped his submarine proposal would get him treatment.

I was called to a Pentagon meeting about time travel. It is well known that light has electric and magnetic fields that are perpendicular to its direction of propagation. All of this comes directly from the Maxwell equations and is consistent with relativity theory that the speed of light cannot be exceeded. The person presenting at the Pentagon said there is a third electromagnetic component in the direction of propagation that was not governed by relativity and could be faster than the speed of light. This meant traveling outside the light cone which meant time travel. The Admiral holding the meeting, apologized, saying he had to leave and would let his staff handle any discussion. I never said a word and the meeting soon ended, a relief for all of the staff.

A scientist who apparently could not get his work funded sent his proposal to his Senator, saying it would benefit the DOD. Of course, the Senator most likely never saw the proposal, but his staff sent it to a Defense liaison, and it eventually made its way to the Chief of Naval Research, my boss, and then to me. I wrote a short reply that we were not interested in the topic. I was asked to rewrite with the explanation that ONR would not go on the record saying there are scientific topics of which we have no interest. I needed to write a complete analysis, complementing the work, but saying it competed with other proposals that ranked higher in the Naval application. That seemed to work, and the matter closed. But some years, ONR received quite a large amount of Congressional plus-up funds, sometimes designating the recipients.

Another time, there was a proposal from a university that came to ONR through the Pentagon. It received a poor review in our Physics Division. Little did we know, but it came back from the Pentagon, saying the Navy liked it more than the Army and Air Force, so ONR will work with the university to get a proposal that you will accept. This was an obvious political pork connection for a building project. The proposal was funded, I remember, for something like $30M.

A professor, whom I funded, had also been the PhD advisor of a grad student who was now his postdoc. The postdoc was Iranian-born. The professor now became the Dean of Science at a different university and the student moved with him, now as a postdoc. All was well, for about 5 years, until the university had a new President who did not support science as strongly as it had been in the past. The Dean/Professor had a strong disagreement with the new university President and together with the postdoc they left for a different university. But trouble followed. Around this time some US based Chinese born scientists were accused, it turned out falsely, of spying for China. So, investigating foreign born scientists was in the air. Well, the Justice Department was asked by the former university to look into the Iranian born post-doc implying he was supplying Iran with classified information. I was contacted for information by the Justice Department. In a phone call, with the local NCIS agent to back me up, I explained that the grant work funded by ONR was 6.1 basic research and all results were published in scientific journals open to the world. I said, if anyone should be investigated, it was the university that was peddling nonsense as retribution for the professor leaving. The postdoc had been at the university for 5 years and it was nonsense for the university to report unfounded suspicions only when he left. All in a day's work for me.

At a meeting at Lake Como, Italy, the American organizer said there are nice places to hike up in the hills. The Italian organizer rushed to the front and said, don't go off the path as there are unexploded ordnance left over from WWI.

In another case, an ONR Program Officer arranged for a proposal to be classified. That Program Officer then retired and joined with the proposer who formed a company. Apparently, it was thought that they could sell the work to industry, with the incentive that ONR thought so highly of the ideas that they classified the proposal. In my view, the proposal read like copying pages from a textbook with nothing new of note. When industry funding didn't materialize, the company wanted to sue ONR, saying they couldn't send the proposal to other funding agencies as it was classified. I wrote a review for our Legal Department, and ONR declassified the proposal. Case closed.

You never know what will pop up on any given day. You might get a request from a Congressional staffer about what the Navy does with lasers. Probably from a 20 something new graduate responding to a question from someone higher up in a Congressional office. We at ONR always tried to be as helpful as possible.

Navy Captain Mark Stoffel and I managed ONR's Counter-Improvised Explosive Device program. The Pentagon set up the Joint IED Defeat Organization (JIEDDO) in 2006. It was founded to expand the Army's Counter-IED Task Force into a joint program and headed by an Army General. Captain Stoffel and I were asked to brief a Congressional staffer, but they asked us more about JIEDDO than our program, and we had no JIEDDO insight.

A small group of us flew out of Andrews Air Force Base to the Naval War College in Newport, Rhode Island. We flew on the Secretary of the Navy's Lear Jet to present lectures at the War College. On the way back to the Newport airfield, one guy in our van said to the Admiral, my wife is a vegetarian, can we stop for dinner to get something to eat with meat? So, we stopped by a fast-food burger place. What a delight to have the Lear Jet and pilots waiting for us and not having us tied to a departure time. Now, late at night, with a van ride back to ONR from Andrews, one guy called his wife. When she answered, I yelled out, "you on the phone pay for your beers". Everyone laughed, even the van driver.

One crazy story was that a woman older than the typical student came to a physics professor's office during official office hours and left quickly. Later, the woman sued the university for sexual harassment by the professor. To make a long story short, the woman and her husband had been setting up professors at several universities with the scheme to win settlements without going to a trial. Apparently, universities were wary of having their endowment vulnerable to a jury judgment and thus settled. Of course, there was never any intention to be a real student, only to extort money on a false charge, and apparently with some success. Sadly, examples such as this one tempered student-faculty interactions.

At ONR, I sometimes had the role of a subject matter expert (SME in DOD lingo) on legal panels with the ONR lawyers for cases when some faculty anonymously accuse others of fraud, waste, and abuse in how federal funds were used. Other times, there were accusations of fabricating experimental results. So, academia was not a white ivory tower but could have stress coming from a number of directions, including adversarial colleagues. Indeed, I thought I had the better deal being at ONR. As a graduate student at the University of Rochester, I attended faculty meetings and served on faculty committees. All interactions were always cordial, and the faculty were supportive of each other and helpful when needed. But maybe not all universities were so blessed.

At the Federal Executive Institute (FEI) in Charlottesville, Virginia, I took a 1 month course to qualify for my SES position. I got to meet new senior executives from several other government agencies. My hotel room had a copy of the Constitution, and *The New York Times* was delivered each morning. Perhaps, the most important lesson that I took away was not to let your job ruin your health. Some people and jobs try. By the way, in 2025, the FEI was abolished as part of executive cost-cutting.

A story that I heard, second hand, was that Ted Litovitz, an expert on the physics of glass, from Catholic University in DC, was seeking funds for an important project. His concept was to store radioactive waste in glass spheres. Supposedly, he approached Tip O'Neill, the Speaker of the House and a Catholic, with the argument that DC does not have Congressional representation, so could O'Neill play that role. Again, the story I heard was that he got the funding for this good and important project, but this would open the gates for universities to go to their Senators and Congressmen for funds, skipping the scientific evaluation process. This supposedly was the start of universities going after pork barrel funding. Yes, over the years, I did later manage Congressional "plus-ups".

I owe a lot to DARPA. The first paper that I published acknowledged DARPA support, and a DARPA grant supported my position at the University of Maryland until I moved to ONR.

From ONR, I had numerous interactions with DARPA. Back when a laptop was a novelty, I was offered one by DARPA in exchange for managing a grant to the Santa Fe Institute (SFI). One day, representing SFI, there came to DARPA the Nobel Laureate Murray Gellman, Senator Jeff Bingham from New Mexico, others, and from the Government side, there included the Head of DARPA, Gary Denman, a military deputy, myself, and others. The SFI presentation was a computer program model that tracked the movement and interaction of three species. The intent was to apply the same methods to track the drug trade. I raised my hand and said I have the computer game SimLife that tracks 500 species and costs about $29.95. I was ignored and never got the laptop or to manage the grant. However, I did later enjoy lecturing and funding a conference at SFI in Santa Fe.

Another time, I came to a math meeting at DARPA. The DARPA manager said that last year they funded wavelets, and this year that was being incorporated to navigate cruise missiles. This left the impression that a math paper method could find actual deployment within a year. When it was my turn to speak, I said ONR funded wavelets years ago, so it was not a new concept, and I said to the mathematicians, no need to buy Jane's book of Fighting Ships to guide your math research or proposal writing. My remarks were not well received.

In 1986, I was on a review panel for DARPA and had a difference of opinion on ranking the proposals. After the awards were made, at a program review meeting, I was introduced (with a Freudian slip) by the DARPA manager as being from Awful Naval Research. I thanked the speaker for an awfully nice introduction.

I heard a DARPA presentation on using detectors to find buried mines by seeking out nitrogen's magnetic quadrupole moment. The concept was born at NRL. The speaker, showed a photo at Eagle Base in Tuzla, Bosnia, and boasted of success in finding all the mines at a testing location. The speaker was a program manager, and in the photo, I recognized a scientist who designed the magnetic detection equipment. I contacted him and he said anyone could find the buried test mines because that was where the grass was worn out. I believe finding buried mines is still a daunting task.

In 2012, I proposed to DARPA a logistics program on supply chains, a combination of mathematical models and economic input. Nothing came of my proposal, but it might have been useful in 2020 when COVID disrupted supply chains. One meeting on logistics opened with the tongue in cheek remark of a ceremony seeing forklifts going by in the missing man formation.

I have reviewed proposals for DARPA, and they have many good scientists and accomplishments, especially for advanced research projects. The above are just some of my quirky interactions from ONR that came up at random times concerning basic research programs and not the advanced research programs that were the bulk of DARPA programs.

A group came to ONR to pitch an idea about a novel internet connection. Jack Kemp, a former Congressional leader, was with the group. Our Admiral asked us to be respectful, as Kemp was an influential person. One of the first slides showed an article by a small local newspaper highlighting the company, and the next slide wrote that electrons move through wires at nearly the speed of light. I raised my hand and objected, saying that an impulse moves rapidly, but the electrons barely move, like the wind blowing through a field of wheat, as a wave moves, but the wheat stays put. Anyway, I added that if the electrons accelerated to a high velocity, they would radiate, and that is the physics behind generating radiation by the free electron laser. The speaker replied, don't worry, that slide was just for Congress. Kemp visibly winced, probably thought what had he gotten himself into. ONR did not fund the project.

One person called me to get our files on the Philadelphia Experiment, where conspiracy buffs claimed the US teleported a destroyer during WWII. I told him ONR was created after WWII, so we did not have any files. He hung up and I guess was looking for a part of the Navy existing during WWII. Anyway, not knowing the location of a ship doesn't mean it was teleported.

Neutrinos are nearly massless and difficult to detect. Can neutrinos from a submarine's nuclear reactor be detected? The proposal came to ONR based on a torsion balance with one arm holding a crystal and the other arm holding a glassy material of the same mass. The answer is no, but that concept was raised, and the submarine community takes seriously any threat to tracking a submarine. An experiment at the NIST reactor in Gaithersburg failed to detect neutrinos, and a DARPA/JASON study found the theoretical foundation not supporting a detection reality. Getting negative results is also sometimes important.

I had a dream where I pictured a house with each room needing something, a new feature, reorganization, or cleaning up. Was told to pick a room and that would be your thesis topic. If you didn't pick a room, then you would need to build your own house. Maybe that's not off the mark on choosing a thesis topic. When I have dreams about thesis topics and ONR, that's a good time to retire.

After the Mast

Richard Dana, after his 2 years before the mast, returned to Harvard as student. After my 40 years before the mast at ONR, maybe I'll look to go sailing

or bird watching

or sit on a beach with my grandchildren, including Leo Kaner, who painted these pictures, not so long ago when he was seven.

Index